May 2002
a find from Emily (coll
to her family.

Other Books by George Sanderlin

ST. JEROME AND THE BIBLE

FIRST AROUND THE WORLD:
A Journal of Magellan's Voyage

EASTWARD TO INDIA:
Vasco da Gama's Voyage

ACROSS THE OCEAN SEA:
A Journal of Columbus's Voyage

1776: *Journals of American Independence*

THE SEA-DRAGON

Journals of Francis Drake's Voyage Around the World

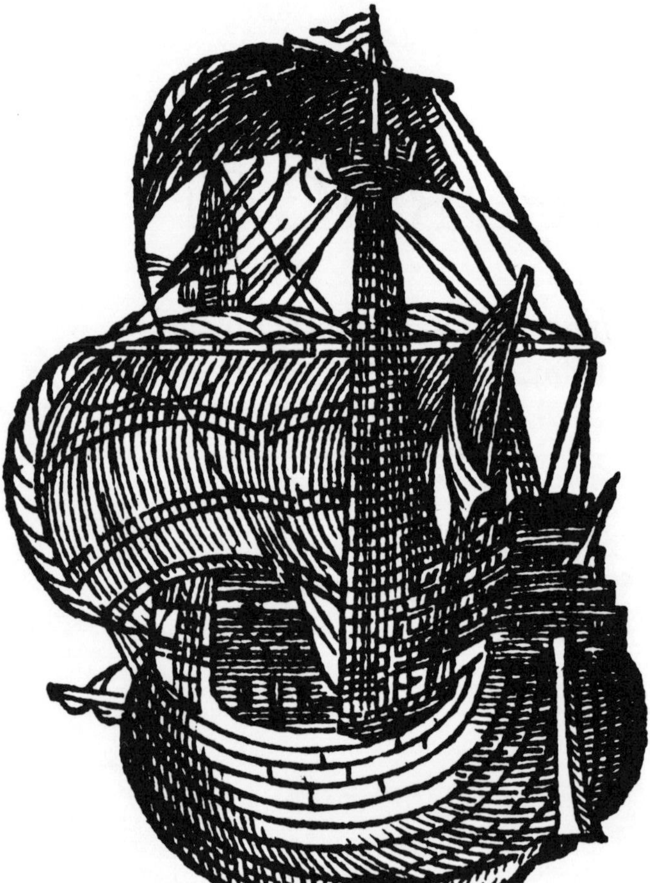

THE

Journals of

George Sanderlin

SEA-DRAGON

Francis Drake's Voyage Around the World

HARPER & ROW, PUBLISHERS
New York, Evanston, and London

Grateful acknowledgment is made for permission to reprint selections from the following books:

DRAKE'S VOYAGES by Kenneth R. Andrews. New York, Charles Scribner's Sons, 1968. Reprinted by permission.

NEW LIGHT ON DRAKE, edited and translated by Zelia Nuttall. London, Cambridge University Press on behalf of The Hakluyt Society, 1914. Reprinted by permission.

SIR FRANCIS DRAKE'S VOYAGE AROUND THE WORLD by Henry R. Wagner. San Francisco, John Howell Books, 1926. Reprinted by permission.

THE WORLD ENCOMPASSED BY SIR FRANCIS DRAKE, edited by W. S. W. Vaux. London, The Hakluyt Society, 1854. Reprinted by permission.

THE SEA-DRAGON: Journals of Francis Drake's
Voyage Around the World

Library of Congress Catalog Card Number: 72–77946

697071727387654321

To My Wife

Contents

LIST OF ILLUSTRATIONS *page x*

AUTHOR'S NOTE *page xi*

INTRODUCTION *page xiii*

1. THE SPANISH MAIN *page 1*

2. A COMMISSION FROM THE QUEEN *page 33*

3. MUTINY AT SAN JULIAN *page 51*

4. THROUGH THE STRAIT *page 77*

5. RAIDING AN EMPIRE *page 101*

6. THE COAST OF CALIFORNIA *page 143*

7. THE WORLD ENCOMPASSED *page 179*

EPILOGUE *page 207*

BIBLIOGRAPHY *page 219*

CAST OF CHARACTERS *page 221*

A TIMETABLE OF EVENTS *page 227*

PICTURE CREDITS *page 233*

INDEX *page 235*

List of Illustrations

JOHN HAWKINS *page xxiii*

SLAVES MINING GOLD AND SILVER *page xxxi*

NOMBRE DE DIOS *page 9*

RAIDING A TOWN ON THE SPANISH MAIN *page 22*

MAP OF DRAKE'S TRIP AROUND THE WORLD *pages 42–43*

QUEEN ELIZABETH I *page 45*

MAP OF SOUTH AMERICA *page 61*

DRAKE AND INDIANS IN ARGENTINA *page 64*

MAP OF THE STRAIT OF MAGELLAN *page 83*

MOCHA ISLAND *page 97*

LLAMAS CARRYING GOLD AND SILVER *page 111*

MINE AT POTOSÍ *page 112*

THE *Golden Hind* CAPTURING THE *Nuestra Señora de la Concepción* *page 133*

DRAKE AND INDIANS IN CALIFORNIA *page 169*

BRASS PLATE FOUND IN CALIFORNIA *page 176*

DRAKE AT TERNATE *page 193*

THE SPANISH ARMADA *page 215*

Author's Note

Most of this book consists of selections from sixteenth- or seventeenth-century works describing the exploits of Sir Francis Drake. Omission of words found in the original source is indicated by a series of dots. A few sentences have been transposed so that the order of events will be clearer, and in four selections the third person "deponent" has been changed to "I" or "me." Words inserted by the editor for additional information or as an explanation are bracketed (e.g., *caracoas* [large open boats]). Punctuation and capitalization used in the original sources have been retained for the most part, but the paragraphing is that of the present editor.

Introduction

One day in 1572, strong English voices rang across the blue water of a secret cove in Panama. Here, in the heart of the mighty Spanish empire, young English seamen manned the sweeps of a low-lying pinnace (a small vessel powered by oars as well as by three lateen sails). At a command from their captain, they drove it between two headlands into the Caribbean.

"God speed our Captain!" said the mariners left behind, watching from beside their huts and the bowling green.

"Port your helm!" ordered Francis Drake.

The triangular sails of the pinnace filled with the offshore breeze. Already to those on the coast the ship was growing smaller. But they could imagine the gleam in Drake's bold eyes as he pointed his vessel northwest toward the trade lanes of Spanish merchantmen lumbering along, unsuspectingly, to Nombre de Dios. Nombre de Dios was a treasure house of gold and silver en route from the New World to Seville.

Now the pinnace was a speck, a blur against the lighter blue of the sky. And now it was gone. There was only the stillness of the empty sea before the little band on shore, and of the rain forest at their backs.

The pinnace disappeared over the horizon. But it would be remembered in history, for the raids Francis Drake and his youths were making against an empire of 160,000 Spaniards and 5,000,000 Indians marked the beginning of a shift of power that would give half the New World to the Anglo-Saxons.

El gran corsario ("the great pirate"), as the Spaniards called Drake, would "singe . . . the King of Spain's beard" again and again. Soon he would sail all the way around the world which the Spaniards and Portuguese had divided between themselves. He would even plan a colony in California for his countrymen.

His daring feats would inspire other Englishmen "to follow his noble steps for gold and silver"—all the way to Jamestown. By trading in the East Indies, he would lead Englishmen to found the East India Company and sail on to India; by plundering the West Indies in the teeth of the Spanish navy, he would prove to them that they could occupy Virginia.

Drake would thus open the windows of insular England upon new prospects—the Far East, the conquest of the oceans, and the settlement of North America.

How did he achieve such success against a supposedly invincible foe? What is known of the early life of "the English Hero"? Who *was* Francis Drake?

HOUSEBOAT ON THE MEDWAY

Francis Drake was born around 1543 in a farm worker's thatched cottage (like Anne Hathaway's). His father, Edmund, was a tenant of a wealthy landowner whose estate had once belonged to English monks. The farm, named Crowndale, was not far from Plymouth, on the English Channel.

Crowndale was in the county of Devon, part of the peninsula of southwest England that thrusts, like a blackboard pointer, into the Atlantic. At the tip of the pointer, the granite cliffs of Land's End face toward the Americas, three thousand miles away. Thus young Drake's earliest years were spent in a land ringed by salt water and swept by Atlantic gales.

If he looked toward the high moorland to the north, there was a gleam above the heather that suggested the sea, and the few trees were torn and twisted by the westerlies. When his father carried him to Plymouth on the lush southern coast where the grass is green all winter, he saw fishing boats, "round" ships from the North Sea, and Mediterranean caravels in the sheltered harbor. Plymouth itself, rising against wooded hills, bore a slight resemblance to the home port of another famous navigator, built on slopes above a crescent harbor—Christopher Columbus's Genoa.

The Drake family remained in Devon only long enough for two or three of Francis's eleven younger brothers to be born. Later, he would remember his father sitting by the fireplace, holding his favorite book in his big calloused hands and reading the passages of Scripture aloud. Edmund Drake's eyes burned with his ardent Protestant faith.

"Make much of the Bible," he advised his sons. "Keep [it] in [your] bosom and feed upon [it]." But Edmund Drake was troubled, for he felt that the Protestants were threatened on every side in Devon by the Catholic majority.

When young Drake was about six years old, in 1549, his father's fears seemed to be justified. Edmund Drake was a supporter of the big Protestant landowners. But most of his fellow tenants, who were Catholic, did not care for either the "new economics" or the "new religion."

The new economics was represented by landowners

who held property once possessed by monasteries. Before the Reformation, many monasteries had farmed out their lands to neighboring yeomen in a kind of sharecropper arrangement; but when King Henry VIII seized the religious houses, he turned their acreage over to favorites and his supporters—"new men" who wanted to get rich quick, and who consequently didn't use, or rent, the land for farming, but for sheep raising. This meant that thousands and thousands of peasants were thrown off their farms, turned into beggars, and cast adrift in the cities.

"Your sheep . . . eat up and swallow down . . . men," Sir Thomas More told these landowners. But English wool, made into rough "white cloth," brought such high prices in the booming Flemish market that "enclosing" the land for grazing continued.

If one dared, he could stage a sixteenth-century "demonstration" by smashing the hedge the new men had raised around the village green—the green where once a poor widow had been free to pasture her cow. If he were lucky, he only had his ears lopped off; if he were unlucky, he was hanged.

In 1549 the government of the boy-king, Edward VI, adopted a new prayer book in English, with Protestant teachings, in place of the Latin missal. The Catholic tenants then added up their economic and religious grievances and arrived at rebellion.

Throughout southwest England great Tudor houses went up in flames. Protestant lords and ladies huddled miserably in woods and caves while peasant bands roamed the fields.

Edmund Drake fled with his family to Plymouth. But then the mayor opened the gates of the city to Catholic rebels, so the sober-faced Protestant refugees rowed hastily out to St. Nicholas Island in the harbor. (This island was later called Drake's Island.) Here they were left undisturbed, but all that winter young Drake's hands were red and raw from the cold, and his stomach ached with hunger. In the spring the government sent ships which transported its "loyal minority" to eastern England, where Protestants were much more numerous.

Here the Drakes established themselves in a new home —a houseboat near the mouth of the Thames. Edmund Drake followed his true vocation and became a chaplain (Bible reader) to the horny-handed sailors of the English navy. Adventurous young Francis Drake was now a minister's son!

Their houseboat was the hulk of a navy "great ship" which was out of commission. Young Drake slept on a cot wedged against the planks of the vessel's side, with the main deck as his roof overhead. A broad roadstead dotted with ships, a watery blue meadow, was his front yard.

This naval base was not actually on the Thames but on the Medway, a river which flows north from Kent, just where that waterway widens to enter the Thames. When young Drake turned his back on the marshy fields and looked out over the roadstead, he had a panoramic view of the Tudor navy.

He saw ships of many shapes and sizes. There were pinnaces (the equivalent of our destroyers), galleasses (small square-rigged vessels with auxiliary oars), and row-barges. But the future safety of England depended on the evolution of two different types of ocean-going vessel: the traditional great ship (the sixteenth-century equivalent to our battleship) and the newer galleon (equivalent to today's cruiser).

A "majesty and terror to the enemy," one authority proudly called the great ship. Those which young Drake stared at were wide deep-bottomed ships of as much as one thousand tons, with low waists and high castles towering at prow and stern. In battle, soldiers in the ship's forecastle and sterncastle could enfilade (rake with gunfire from end to end) with their muskets any boarders who gained the waist of the ship.

On the other hand, this same low waist was the camp and battlefield of the English force. Soldiers stationed here kept up a hot fire themselves as the great ship maneuvered clumsily toward the enemy. Seamen cast out grappling irons. Then the soldiers streamed up ladders to prow or stern and across gangplanks to engage in a hand-to-hand melee in the enemy ship.

Naval battles between great ships consisted of a number of isolated fights like this. The ships were simply floating islands which grappled as soon as possible so that soldiers could charge back and forth across the decks as though they were on land. Spaniards favored this kind of sea action.

But the great ship, although it could transport formidable numbers of fighting men, had some disadvantages. It was slow and awkward to sail; its heavy castles strained the timbers of the hull and caused leaks; and the stone cannon balls of its bombards, although fired with a roar like the crack of doom, were as harmless as popcorn to vessels a few hundred yards away.

These disadvantages were overcome in the galleon, or "race ship," of about four hundred tons—a new type developed simultaneously in the north and in the Mediterranean. Its graceful silhouette contrasted with the sky-scraper effect of the great ship. Its castles were much lower, and the open waist found in the great ship was decked over. The galleon was also proportionately longer and narrower than the great ship—three times as long as it was wide.

The armament of the English galleon consisted of twenty or thirty lighter guns whose shot carried twice as far as the balls of the bombards. Spanish galleons, however, carried heavy short-range cannon.

The invention of the galleon led to a change in naval tactics in which England was the pioneer. The English were the first to take the management of sea fights away from the soldier and turn it over to the mariner.

The galleon was swift and maneuverable, but it could not sail close to a great ship and "slug it out," nor was it well-adapted to boarding. Therefore English captains—who commanded their galleons, unlike Spanish captains who were subordinate to the ranking soldier aboard—fought at long range with their artillery. They tried to sink the enemy's ships, not to board and kill or capture his soldiers. Thanks to these tactics, England was to withstand the Spanish Armada in 1588 and win control of the seas of the world.

But that was in the future. There was only a hint of this development in the ships young Drake admired. The navy before him was a navy in decay. It consisted of only twenty-six ships, compared to fifty-three under warlike Henry VIII. New vessels had been so badly built, owing to graft, that several were sold to private individuals for a few hundred dollars each. And old-fashioned oared galleys, which would be of little use in the ocean conflict with Spain, drilled like college racing shells on the sheltered estuary. (The Mediterranean galley had dominated naval warfare for two thousand years.)

Probably young Drake himself learned how to handle an oar before he learned to sail, as he and his brothers rowed a small boat around the bay, fished between the great ships, or explored coves and swamps along the low shore.

Francis Drake did not grow tall, but his shoulders and

chest thickened. His brown hair stood up from his forehead in a kind of crew cut, and his round eyes always sparkled at any chance for adventure.

In a crisis, young Drake made decisions quickly and well. He had few worries, perhaps because he had acquired from his preacher father a firm belief in a God who would somehow always look out for Francis Drake and Francis Drake's friends.

"Make much of the Bible," young Drake's father had advised him. With almost no formal education, Drake would always cling to that one book.

But suppose Englishmen who sought to "feed upon it" and decide religious questions for themselves were persecuted for their faith?

Edward VI had died young, and his Protestant advisers had scattered. Edward's sister, Mary Tudor, succeeded him—and she was a fervent Catholic. This serious-minded, unhappy queen tried to restore Catholicism in England first by forgiveness and patience—but later she was burning Protestants at the stake!

Young Drake heard with righteous indignation the report of Protestant Hugh Latimer's words to his fellow martyr Nicholas Ridley just before the fagots were lit: "Be of good comfort, Master Ridley, and play the man. We shall this day light such a candle, by God's grace, in England, as I trust shall never be put out!"

Already Drake had a quick, shrewd eye for getting ahead in this world. To it, he now added a stern determination that his England should never be Papist. When later in the century Catholics were fined for attending Mass and their priests were martyred, Drake considered it necessary for the defense of England against a foreign foe—the Pope, or Spain. He would look on his piracies as crusades against these same hated Papists.

Not long after Mary came to the throne, young Drake's life on the Medway came to an end. He was probably not quite in his teens, but times had become hard for his father, who now apprenticed Francis to the master of a bark which traded along the east coast of England and across the North Sea to Holland.

This bark was probably a descendant of the medieval "Hansa cog," a small stubby vessel, square-sailed like most northern ships and clinker-built (with planks overlapping one another). Here young Drake learned to navigate and to endure the bitter winter weather described by an Anglo-Saxon poet centuries before:

> Beholding gray stretches of tossing sea,
> Sea-birds bathing, with wings outspread,
> While hail-storms darken, and driving snow.

He saw English fishing boats bound for Dogger Bank, which runs east and west across the North Sea. Fishing was so profitable to England that Queen Elizabeth required her Protestant countrymen to eat fish twice a week even though this might remind them of the old Catholic custom of abstinence on Friday. Sacrifices for one's religion were good, but they need not cause the loss of hard cash!

Drake also saw large convoys from Seville, beating up the English Channel, bearing provisions and Spanish pikemen—the best in Europe—to the restless Low Countries. In 1516 the Netherlands and Spain had been united under the rule of the Hapsburg Prince Charles, later the Emperor Charles V. But now in the early 1560's, inflamed by the Protestant teachings of John Calvin, the Netherlands was on the brink of revolt against Philip II and Spain.

Great crimson crosses were blazoned on the sails of the supply ships and armed galleons passing through the Channel; gold from the Spanish Americas was part of their ballast.

Indeed the riches of the Indies carried in Spanish ships were fabulous. Indian masks of thin beaten gold, gold nuggets as large as eggs, gold and silver bracelets and rings, a sun made of gold and a moon of silver, emeralds as long as a man's finger, rubies, and other gems—not to mention the gold and silver of the mines, like Zacatecas in Mexico and Potosí in Bolivia.

In a century and a half, 200 tons of gold and 18,600 tons of silver would pour into Spain. Francisco Pizarro discovered a treasure room of the Incas in Peru, 22 feet by 17 feet, filled with silver and gold. Hernando Cortes seized the wealth of Montezuma in Mexico. From 1545 on, the mines of Potosí, a barren, white-capped cone rising fifteen thousand feet into the sky, were an inexhaustible source of shining silver.

As young Drake watched this pageant of imperial Catholic power, he may have wondered whether the few great ships on the sleepy Medway could ever protect England against it. How delightful it would be though to divert some of that treasure into the pockets of a true Englishman—such as Francis Drake!

After only a few years, young Drake's master died and left his ship to Francis. But, says an Elizabethan historian, Drake "soon grew weary of his bark which . . . crept along the shore [of eastern England]" like a coal barge. When he was eighteen, Drake sold his bark and signed as third mate on one of the vessels of John Hawkins of Plymouth, a distant cousin of his.

If he wanted adventure, the impetuous Drake had come to the right man. John Hawkins had killed the town bully, a barber, with one blow when he was twenty and had become an independent merchant shortly after. Now he was

John Hawkins, a distant cousin of Francis Drake. He made Drake captain of the Judith, *one of the six ships he took to the Caribbean in 1567.*

the most important shipping tycoon in England. With his headquarters in London he was, in effect, the head of the English merchant marine.

He was also a handsome, captivating man who could have sold snowshoes in the Sahara. Now he had a better idea: He would sell Negro slaves at discount prices along the Spanish Main (the West Indies and the Caribbean coasts of South and Central America). He would thus undercut Spanish traders, who had to buy an expensive license from their government to deal in slaves. That it was strictly against Spanish law for outsiders to trade in Spain's New World empire seemed only a minor obstacle to John Hawkins. Drake took to him immediately.

After making several voyages under other captains, Drake awoke one morning in 1567 to discover that John Hawkins had selected him for the third and most crucial expedition to the Caribbean.

At age twenty-four, Drake was to command a ship in the heavily armed squadron that was about to plunge into the Atlantic and head southwest—toward a major international crisis.

ENGLAND AND SPAIN

John Hawkins's proposed expedition had made Lord Burleigh, England's prime minister, grow pale. Lord Burleigh—"of all men of genius . . . the most a drudge," according to a contemporary—was a true conservative. He wished to make changes slowly and to continue whatever had been successful in the past. Since one of England's most successful political and commercial relationships had been with Spain, Burleigh had no wish to destroy it, especially for the sake of profits on a handful of slaves.

He tried to communicate his fears to Queen Elizabeth,

but "Gloriana," stealing a glance at herself in a full-length mirror, was noncommittal. The daughter of Henry VIII and Anne Boleyn listened to Burleigh's arguments without either agreeing or disagreeing.

She was annoyed that Burleigh had said nothing about the snowy Flemish lace at her throat or the new diadem that brought out the red-gold in her hair. Reckless, dashing Leicester, spokesman for the hawkish anti-Spanish party, would at least have done that!

On the other hand, Burleigh's reasoning was as solid as his figure. Hawkins's venture was dangerous; it might bring needed gold into impoverished England, but it might also rouse Elizabeth's brother-in-law, Philip II of Spain, to take action. Philip was a pedantic bureaucrat who moved slowly—but he did move. Elizabeth didn't want trouble. . . .

In the end, the Queen did nothing. Perhaps the question would settle itself. Queen Elizabeth preferred problems that solved themselves.

England and Spain *had* been friends, however, for a long time. Even in the Middle Ages, English cogs carried tin, lead, hides, and grain to Spanish ports in the Bay of Biscay. But the most valuable English exports went to the Netherlands: raw wool in the fourteenth century, unfinished cloth in the fifteenth and sixteenth centuries. At the beginning of the sixteenth century, members of the Merchant Adventurers company, exporters of the unfinished "white cloth," rejoiced through fifty boom years as their trade with the Netherlands tripled.

No wonder Lord Burleigh had wrung his hands at the thought of endangering English interests in the Low Countries. In addition, France, England's traditional enemy, was conveniently sandwiched between—encircled by—the allies, England and Spain.

Two revolutionary events near the beginning of the sixteenth century lit a fuse which led ultimately to the shattering of the alliance. One was Columbus's discovery of America; the other was the Protestant Reformation.

Before 1492 the Mediterranean cities, especially Venice and Genoa, had been the middlemen of Europe, importing spices, silks, and other goods from the East and exporting them to northern Europe. Columbus's discovery offered the Atlantic nations the opportunity to surpass Italy in wealth and power. Instead of being on the fringe of the known world, Spain, Portugal, France, and England found themselves at its commercial center. Rivalries were bound to develop.

First, Pope Alexander VI (1493) and later the rulers of Spain and Portugal (1494) prevented war by dividing the New World between Spain and Portugal. Thus far, only Spain and Portugal had explored or claimed the newly-found lands. But by the 1520's, Francis I of France was tartly asking to be shown "the will of Adam constituting them [Spain and Portugal] his universal heirs"—and sending his privateers to raid the Caribbean.

More slowly, England turned to the sea. John Cabot's voyage to North America (1497) was not followed up. In the early sixteenth century, however, English merchants living in Seville were aware of Spanish exploration and conquests. One of them, Robert Thorne, urged Englishmen to get into the game by voyaging to the Pacific by way of the North Pole!

Sir Thomas More showed his interest in the New World by locating his imaginary "Utopia" (1516) in South America. In the 1530's, William Hawkins, John Hawkins's father, traded in dyestuffs along the coast of Brazil. But England did not really make its presence felt overseas until after 1550; then, led by Drake, England was to become Spain's chief rival in the Americas.

As for the Protestant Reformation, when Martin Luther nailed his attack on indulgences to the church door in Wittenberg (1517), his hammer echoed from the Baltic to Gibraltar. By 1520, this strong-minded Augustinian monk, who worried about his salvation with a peasant's earnestness and who was outraged by corruption in the Church, had moved to a truly revolutionary position.

He argued that man was saved by God's grace alone. Man did not need papal indulgences, which the common people had been tricked into considering passes to Heaven. Man did not even need good works inspired by the Church or by his own will.

Every man was to become his own priest. He would no longer require the services of the established clergy for his salvation; he was not bound by clerical law, courts, officials, custom, or authority.

Humanists like Thomas More and his friend Erasmus hoped that there could be a compromise between Lutherans and Catholics. Instead, positions hardened, and Europe divided into two armed camps—in the name of Christianity—with England and Spain on opposite sides.

Thus when young Drake thought of Spaniards, he thought of Papists, of the Spanish Inquisition and its victims—some of them English—paraded through the streets, wearing pictures of Hell Gate as wide as baseball chest protectors on their breasts and tall dunce caps on their heads. They had been sentenced at an *auto-de-fé* ("judicial decree concerning the faith") to make arduous pilgrimages, to imprisonment, or to burning outside the city.

On the other hand, when a Spaniard spoke of Englishmen, he imagined Lutherans in the Caribbean—drunken soldiers looting a Spanish church, smashing the statues of the saints, stealing gold and silver candlesticks from the altar, drinking wine from the chalice, and beating or hanging Franciscan friars.

But the growing commercial rivalry between England and Spain had perhaps even more to do with ending their friendship than the religious conflict did. The Antwerp market for English "white cloth" declined because of depression and the revolt of the Netherlands against Spain in 1568. English merchants then became all the more determined to find new markets overseas, and this increased the likelihood of a collision with Spain.

However, not all English merchants and statesmen advocated challenging Spain in the New World as Drake did. One group wished to discover new lands that might equal or surpass the Americas in wealth. But where would one look for these other windfalls? One could not sail either west or south across the Atlantic without bumping into a Spanish or Portuguese "No Trespassing" sign.

Almost the only route left was northeast in search of a short cut to India over the top of Russia. John Dee, an eccentric geographer and magician who conversed with spirits, championed this route. In 1552, Hugh Willoughby and Richard Chancellor attempted it, and Chancellor wintered in Archangel, Russia. Their voyage led to the formation of the profitable Muscovy Company, in 1555, for trade with Russia—but neither they nor their followers ever reached the Ganges by way of the Arctic ice.

John Dee rubbed his "long beard as white as milk" and made another proposal: Let the mariners search below the Strait of Magellan for Terra Australis ("The Southern Land"), which was supposed to contain the riches of the biblical Ophir. John Dee and other geographers believed this southern continent had to exist to counterbalance

Europe and Asia in the northern half of the world. Like a weight at the bottom of an inflated rubber clown, it was supposed to keep the sphere from toppling over.

Although much discussed, this quest was not undertaken until Drake set out in 1577 on his voyage around the world. Of course, the Terra Australis of gold and spices was never found. But it did not disappear from maps, or from man's imagination, until the eighteenth century. The mythical continent finally bequeathed its name to Australia.

The English search for a northwest passage to the Pacific, a kind of imaginary Panama Canal through Canada, is better known. It was advocated by Humphrey Gilbert (1566), and unsuccessful attempts to find it were made by Martin Frobisher (1576–1578), John Davis (1585–1587), and many others. Richard Hakluyt, the chief Elizabethan writer on geography, described these voyages in his *Principal Navigations* (1589).

In the end, England's breakthrough to markets and colonies would not come from the discovery of new lands or new routes to the East, but from the incursions of Drake, Hawkins, and others into territory already conquered or claimed by Spain. First in the Caribbean, the vestibule of the Americas, later as far away as Peru, they would force their goods on the Spaniards, or shamelessly seize Spanish vessels and Spanish treasure.

Hawkins and Drake began the aggression against Spain by smuggling slaves into settlements in the Caribbean.

A TROUBLESOME VOYAGE

The word *slave* is derived from *Slav*—because in the great slave market at Venice during the Middle Ages, Slavs made up the majority of the unfortunates. Nevertheless, slavery was not widespread in medieval Europe, having been replaced by serfdom on the feudal estates.

It was not until the fifteenth century that the Portuguese began capturing or buying Berbers and Negroes in West Africa and selling them into slavery in Portugal. In the sixteenth century, Negro slaves were especially in demand in the New World because they could endure hard physical labor better than the native Indians.

By that time this cruel form of colonialism was taken for granted by most Europeans. As Drake was rowed out to his ship, the *Judith*, in Plymouth harbor, he felt no humanitarian scruples; he had no doubt of the white man's God-given right to exploit the black man. If all went well on this important slaving expedition, he, a poor minister's son, might become rich at age twenty-four!

All had gone remarkably well on Hawkins's two previous expeditions in 1562 and 1564. Queen Elizabeth, a cautious but canny businesswoman, even put her concern about Philip II to sleep long enough to become an investor —she was lending her great ship, the *Jesus of Lubeck*, for this third venture.

It was true that Spanish officials had emphatically rejected the offer from Plymouth's traveling salesman to police their own Caribbean for them in return for trading privileges there. It was also true that Pedro Menendez, the Spanish naval genius, had just organized the new swift "galleons of the Indian guard" to protect Spanish ships and to clear the Caribbean of intruders. It made no difference to the enterprising Hawkins—and Drake laughed at the idea of Spanish Papists holding their own with Englishmen on the high seas.

Drake also tried to smile at the evil omens noted by the crews—omens of what Hawkins was later to call his

Slaves from Africa were forced to mine silver and gold for the Spanish in the New World.

"troublesome . . . voyage." There were enough gloomy signs among the six ships for a cohort of witches!

First, on the *Jesus of Lubeck* some heavy gear broke loose and killed a young girl visitor. Somewhat later as the ships waited in Plymouth harbor, seven Spanish vessels entered seeking shelter from a storm, but they did not observe custom and lower their flags to the Queen's ship until John Hawkins fired on them—an international incident which led to a long unpleasant discussion between the Spanish ambassador and Elizabeth.

At last the six ships—the *Jesus of Lubeck* and the *Minion*, both great ships, accompanied by the *William and John*, the *Swallow*, the *Angel*, and Drake's *Judith*—got under way October 2, 1567. Then on the first leg of the voyage, a tempest nearly sank the ancient *Jesus*, and one of Hawkins's officers went berserk and stabbed him in the face.

In spite of all, Drake could see John Hawkins across the waves, tramping the quarterdeck, unperturbed. Hawkins's magnificent crimson velvet breeches and scarlet leather jacket trimmed with silver braid brightened the dull days. Every evening the crew knelt with him at the mainmast—they were flogged if they didn't—and he led them in reciting the Psalms, the Lord's Prayer, and the Creed. Then he retired to his cabin to make plans for capturing the Negroes whose sale would enrich him and his backers.

Drake may have fought in the raids that Hawkins ordered when he reached West Africa. Twice, his men, seeking to kidnap Negroes, were driven off by the Negroes and the Portuguese. But Hawkins picked up 150 slaves in the holds of the Portuguese caravels he captured; then he took part in a local war and made up the rest of a cargo of about 470 slaves from the captives his men seized or had been given by allies. The fleet crossed the Atlantic and arrived in the West Indies in March, 1568.

Here, at first, even the impetuous Drake had to admit that John Hawkins's "soft sell" worked. Hawkins would address a letter to the governor of an island, explaining that bad weather or a shortage of supplies had forced him to stop there. All he wanted was food and water, for which he would pay well, and, as he just happened to have a cargo of Negro slaves aboard, to be given a temporary license to sell a few of them so he could meet expenses.

Hawkins always included a tactful reference to Philip II as "my old master" (Philip had been King of England while married to Mary Tudor). And somehow Hawkins managed to give the impression that he was Catholic—although, of course, he could not say much about that among English heretics. He had, he concluded, some splendid personal gifts for his Excellency, the Governor.

The governors of the first two places he came to, the island of Margarita, and Borburata in South America, looked at these gracious letters and then out across the water into the mouths of English cannon—and they gave way. But at his third port of call, Rio de la Hacha, also on the coast of South America, there was trouble.

Drake had been ordered to go ahead of the fleet with two ships and reconnoiter. It was a pleasing assignment—Drake was spoiling for action, and he had a score to settle with the chief official of Rio de la Hacha, the treasurer. On Drake's one earlier voyage to the Caribbean, as a mate—a voyage about which little is known—this treasurer had tricked Captain John Lovel and Drake out of some slaves. (Drake never forgot or forgave a blow to the pocketbook.) Just let the treasurer, Miguel de Castellanos, try that again!

Drake's salesmanship was different from Hawkins's.

When shore batteries at Rio de la Hacha fired on his two ships, Drake cried, "I will remain here in despite of the Spaniards and their shot!" And he put a cannon ball through the house of Miguel de Castellanos.

Then for five days he blockaded the town, and he seized a government dispatch boat—an act of war at a time when Spain and England were officially at peace. He neglected to address any complimentary letters to the stubborn treasurer.

It was too late for letters when Hawkins arrived. To obtain water to keep his slaves alive, Hawkins had to land a force of two hundred men and capture the town. Even then, the Spaniards fled to the jungle and held out for a few days under the courageous Castellanos—until a Negro slave escaping from the Spaniards revealed to Hawkins where Castellanos had hidden the town's treasure.

"John Hawkins . . . is such a man that any man talking with him hath no power to deny him anything he doth request," said Castellanos bitterly, now forced to permit Hawkins to sell his slaves. According to Spanish sources, one of the slaves Hawkins sold was the escapee who had told him about the treasure!

At Cartagena, capital of the Spanish Main, things were still worse. The Spaniards here absolutely refused to deal with Hawkins, and even Drake would have hesitated to storm this powerfully defended city.

So Hawkins turned back toward Europe. The fleet was sailing past the west end of Cuba, headed toward the Florida Straits and the Gulf Stream when a hurricane ("furicane," Hawkins called it) struck.

For four days in August, 1568, the tropical storm lashed the squadron. It drove the ships apart. It ripped the high castles off the *Jesus of Lubeck* and smashed that vessel's rudder. Darkness and sheets of torrential rain hid sky and sea. When the weather finally cleared and the ships limped back together—all but the *William and John*, which was headed for Ireland—Hawkins found that Elizabeth's *Jesus of Lubeck* leaked like a sieve. Fish were swimming around in her hold!

He could have abandoned her, and the loss would have been the Queen's—but John Hawkins, like Drake, knew how it felt to lose money. He determined to repair the *Jesus*. A northwest gale left him little choice of ports to sail to. Driven into the Gulf of Mexico, he learned from a captured Spanish pilot that San Juan de Ulua, Mexico, was the only port he could make. He went there.

Next to Seville itself, Hawkins probably could not have chosen a port where he would be less welcome. In the first place, the incident of his having fired on Philip II's ships in Plymouth was probably known there. In the second place, the harbor of San Juan de Ulua, a mere roadstead sheltered from northers by a low sandy island, was already occupied —by eight Spanish vessels loaded with silver worth two million dollars.

Drake laughed at the sight of nervous Spaniards fleeing from the island to the mainland when they learned that the strange black ships were English. But his round eyes became rounder at rumors of the silver on the Spanish vessels.

Hawkins, as usual, got into the harbor by a stratagem. On sighting the port on September 15, 1568, he had ordered his captains to take down all flags bearing England's distinctive red cross of St. George. He had learned from captured Spanish pilots that San Juan de Ulua was awaiting the arrival of a "Plate Fleet" (a squadron from Spain, escorted by armed galleons, which would carry the treasure in the harbor back to Seville).

The Spaniards of San Juan de Ulua then mistook Hawkins's ships for the Plate Fleet. On September 16, the King's treasurer and the deputy governor of nearby Vera-cruz came out in launches to welcome him—and Hawkins politely detained them while he sailed his squadron in, past the island batteries which commanded the harbor entrance. Then he promised not to touch their treasure—he only wanted to supply and repair his ships. As a safeguard, he persuaded his hostages to turn the island batteries over to his men.

Unfortunately the next day the real Plate Fleet came over the horizon. Hawkins's and Drake's superstitious sea-men gloomily noted that it was Friday and counted the Spanish ships—thirteen.

On one of these galleons was the new Viceroy of Mexico, Don Martin Enriquez. With Hawkins's men now manning the island batteries, the proud viceroy made the unhappy discovery that he had to ask permission from an English heretic to go into his own house.

"I am a Viceroy and have one thousand men, and there-fore I will come in!" said Don Martin Enriquez in a mes-sage to Hawkins.

"I represent my Queen's person and [therefore] I am a Viceroy as well as he," retorted the irrepressible Hawkins. "And if he has a thousand men, my powder and shot will take the better place!"

In the end, September 20, the viceroy had to humble himself and accept John Hawkins's terms: Hawkins could repair his ships, sell his goods, and hold the island and its batteries. The viceroy did not dare remain in the open sea for fear of a September hurricane and so entered the har-bor. On the other hand, since England and Spain were not at war, Hawkins had to tread the narrow line between protecting his men and creating another incident which might cause Elizabeth to punish him severely.

But the furious Enriquez had already summoned soldiers from nearby Veracruz. They crept into an empty hulk next to the *Minion* at night on September 22—Spanish and English ships had to be berthed only twenty yards apart, fastened by iron chains to a quay on the landward side of the island.

The following day, September 23, many "sight-seeing" Spaniards came to the island. Worse, a number of Spaniards were seen on the hulk. Hawkins sent the master of the *Jesus of Lubeck* to Don Martin to protest—whereupon the Spaniards promptly seized him. Drake, said to have been on shore at the time, heard a trumpet call sounded ominously on a Spanish flagship, and then the clash of swords and muskets firing.

"Treachery!"

Drake leaped up. He raced along the quay, caught a hawser, and climbed up into the *Judith*. Only two other English seamen ashore escaped capture or death.

Musket fire, cries, acrid smoke drifting over the ships, axes smashing into the iron chains—all was confusion. Spanish pikemen swarmed from the "empty" hulk into the *Minion*.

"God and St. George!" Drake heard Hawkins cry. "Upon those traitorous villains and rescue the *Minion*!"

Panting from his run, Drake shouted orders. He got the *Judith* under way. Meanwhile, Hawkins led his men from the *Jesus of Lubeck* into the *Minion* and drove the Spaniards off.

Moorings were cut. English and Spanish ships drifted into the harbor, firing at each other at close range.

The two largest Spanish ships burst into flames. One sank, but it did not disappear from sight because the water was so shallow. Drake headed for the harbor entrance, firing all his guns.

But the Spaniards now held the island, the key to the battle. They retook the batteries and began firing. Their shot missed the *Judith* but sank the *Angel* and disabled the *Swallow*. A cannon ball knocked a tankard of beer out of Hawkins's hand before he could drink it, splashing his scarlet jacket.

"Fear not, for God who hath delivered me from the shot will also deliver us from these traitors and villains!" roared Hawkins.

Now only the *Jesus of Lubeck* and the *Minion* were left close to the island. Hawkins tried to transfer the $130,000 worth of gold pesos received in trade for his slaves from the damaged *Jesus* to the *Minion*.

But as Drake watched, furious and helpless, from the harbor entrance, the Spaniards launched two fire ships, blazing like forest conflagrations, straight at the *Jesus of Lubeck*.

"Fear not!" shouted Hawkins. "Fear not—"

But no one was listening. Exhausted after hours of fighting, the English seamen had finally panicked. They ran past Hawkins and leaped across to the *Minion*. Reluctantly, the last to leave Elizabeth's ship, Hawkins joined them.

It was growing dark. Hawkins guided the *Minion* safely past the island guns, out into the open sea. The battered Spanish ships made no pursuit. That night a stunned Drake sailed off for home in the *Judith* without notifying Hawkins; the next day the *Minion* followed.

John Hawkins's troublesome third voyage had indeed ended in disaster. Barely a fifth of the hopeful company which had sailed from Plymouth a year earlier returned. Spanish or Portuguese bullets, scurvy, poisoned arrows of Negroes in Africa, fever, drowning, even—for a handful—the Inquisition in Seville, took them off. Francis

Drake lost the small savings he had invested in the scheme. Philip II's ambassador in London lost his temper. Cautious Elizabeth lost her expensive ship after all.

But a turning point in English-Spanish relations had been reached, a point of no return which would give young Drake his opportunity—when he recovered—for revenge, fame, and fortune. After San Juan de Ulua, according to an Elizabethan historian, "the military and seafaring men all over England fretted and demanded war against the Spaniards." The Pearl Harbor of the Spanish Main had taken place.

Queen Elizabeth would now be more willing than ever to look the other way if Drake wished to "singe . . . the King of Spain's beard." His commando approach would replace Hawkins's politicking. He would find backers for voyages of thinly—or not so thinly—veiled piracy. And when he saw, first of Englishmen, a strange romantic South Sea from a mountaintop and felt his ambition stir to sail on it—for Francis Drake's ambition encompassed the whole world—he could follow that ambition, in the *Golden Hind*, from Plymouth to the Spice Islands and home again.

This book tells the story of Francis Drake, *el gran corsario* and scout of the Anglo-Saxon empire. It follows him from his return to the blue waters of the Caribbean on a daring raid against the empire which had wronged him, through the intrigues of Elizabeth's court, to his circumnavigation of the globe—a voyage of adventure, danger, and prophecy. The denouement is the defeat of the Spanish Armada (1588), which gave the English-speaking peoples command of the sea lanes that lead to the New World.

Chapter One

THE SPANISH MAIN

The star of your good fortune
bore your boldness to Panama.
—LOPE DE VEGA

One

A SECRET HARBOR

"He [is] Low of stature, of strong limbs, broad Breasted, round headed, brown hair, full Bearded, his eyes round, Large and clear, well favoured, fair, and of a cheerful countenance. . . . In his imperfections he [is] Ambitious for Honor, Unconstant in amity, Greatly affected to Popularity."

Thus an Elizabethan historian, John Stow, described Francis Drake, and thus he probably appeared on May 24, 1572, as he stood on his quarterdeck on the *Pascha* (70 tons) and saw the hills of Plymouth fall away behind him. The choppy Channel waves struck his vessel, and the tiny *Swan* (25 tons) following. Seventy-three men and boys, only one over thirty years of age, worked the sails and shared Drake's dreams of plunder.

Two were Drake's brothers—the too-daring John Drake, captain of the *Swan*, and young Joseph, a seaman on the same ship. Another leader was John Oxenham, first mate of the *Swan*, a reckless adventurer who hated Papists even more violently than Drake did.

Four years had passed since the treacherous attack upon Hawkins and Drake at San Juan de Ulua, and in that time English opinion had turned against Spain. The Spanish ambassador was implicated in the just-discovered Ridolfi

Plot to assassinate Queen Elizabeth. Englishmen thought, mistakenly, that Spain had backed the Pope's excommunication of Elizabeth in 1570. John Hawkins and other wealthy merchants met no objection from court when they prepared the expedition on which Drake was now embarking—an expedition not for armed trade, but for piracy.

Drake had married Mary Newman in 1569 and needed more money than ever. He hoped to seize the Spanish treasure which was transported every year from Panama, on the Pacific coast of the Isthmus, to Nombre de Dios, on the Atlantic coast, and from there shipped in the annual fleet (*flota*) to Seville. In 1570 and 1571, he had made voyages to the Isthmus "to gain . . . intelligence" of the handling of the treasure. At that time he discovered a hidden harbor on the Isthmus, near South America. He named it Port Pheasant and buried supplies there.

Now, after a fast crossing of the Atlantic in twenty-five days, Drake found himself again in his favorite hunting ground, the Spanish Main—the Caribbean with its islands and surrounding coasts. He sailed straight to his secret harbor on the Isthmus.

On July 12, 1572, sheets were slackened, sails came down, and the two ships came about at a lonely spot on the shore of eastern Panama. Before them, between two headlands, a silent bay opened out—a sandy beach—a wall of dense green forest.

"Port Pheasant!" Drake clapped his hands joyfully. "Man the ship's boat!"

Seamen lowered the *Pascha*'s longboat and clambered into it. Drake leaped down beside them. Oars splashed, and the longboat was rowed swiftly toward land—and toward a chilling surprise.

What Drake found is related in the following selection.

It is from Philip Nichols's *Sir Francis Drake Revived* (1625), a work dedicated earlier by Drake himself to Queen Elizabeth and said to have been revised by him, but printed in 1625 to arouse enthusiasm for a war then being carried on with Spain.

On Whitsunday Eve, being the 24th of May, in the year 1572, Captain DRAKE in the *Pascha* of Plymouth of 70 tons, his admiral [flagship]; with the *Swan* of . . . 25 tons, his vice-admiral, in which his brother JOHN DRAKE was Captain (having in both of them, of men and boys seventy-three, all voluntarily assembled; of which the eldest was fifty, all the rest under thirty . . . and . . . having three dainty pinnaces . . . taken asunder . . . and stowed aboard . . .) set sail, from out of the Sound of Plymouth, with intent to land at Nombre de Dios.

The wind continued prosperous and favourable at north-east, and gave us a very good passage . . . so that . . . we never struck sail, nor came to anchor . . . until twenty-five days after; when (28th June) we had sight of the island of Guadaloupe, one of the islands of the West Indies, goodly high land.

The next morning (29th June), we entered between Dominica and Guadaloupe, where we descried two canoes coming from a rocky island . . . which usually repair thither to fish. . . . We landed on the south side of it, remaining there three days to refresh our men; and to water our ships out of one of those goodly rivers, which fall down off the mountain. . . .

The third day after (1st July), about three in the afternoon, we set sail from thence, toward the continent of *Terra Firma* [South America]. And the fifth day after (6th July), we had sight of the high land of Santa Marta [Colombia]; but came not near the shore by ten leagues.

But thence directed our course, for a place called by us, Port Pheasant; for that our Captain had so named it in his former voyage, by reason of the great store of those goodly fowls, which he and his company did then daily kill and feed on, in that place. . . .

Within six days . . . we arrived (12th July) at Port Pheasant, which is a fine round bay, of very safe harbour for all winds, lying between two high points . . . having ten or twelve fathoms of water . . . full of good fish; the soil also very fruitful, which may appear by this, that our Captain having been in this place [in July, 1571] . . . and having rid the place with many alleys and paths made; yet now all was . . . overgrown. . . .

Our Captain . . . was on his way, with intent to have gone aland with some few only in his company, because he knew there dwelt no Spaniards within thirty-five leagues of that place. Santiago de Tolou being the nearest to the eastward, and Nombre de Dios to the westwards, where any of that nation dwelt.

But as we were rowing ashore, we saw a smoke in the woods, even near the place which our Captain had aforetime frequented; therefore . . . he caused his other boat also to be manned. . . .

When we landed, we found by evident marks, that there had been lately there, a certain Englishman of Plymouth, called JOHN GARRET, who . . . had now left a plate of lead, nailed fast to a mighty great tree . . . on which were engraven these words. . . .

CAPTAIN DRAKE!

IF YOU FORTUNE TO COME TO THIS PORT, MAKE HASTE AWAY! FOR THE SPANIARDS WHICH YOU HAD WITH YOU HERE, THE LAST YEAR, HAVE BEWRAYED [REVEALED] THIS PLACE, AND TAKEN AWAY ALL THAT YOU LEFT HERE.

I DEPART FROM HENCE, THIS PRESENT 7TH OF JULY, 1572.

YOUR VERY LOVING FRIEND,

JOHN GARRET.

SEVENTY-FOUR ENGLISHMEN CAPTURE A CITY

John Garret was an old friend of Drake's who had been an officer in John Hawkins's fleet. The smoke at Port Pheasant was from his fire—he had just left.

But Spaniards or no Spaniards, Drake would remain at Port Pheasant at least long enough to put together the three pinnaces whose prefabricated parts he carried. He needed these swift, low ships to catch Spanish merchantmen. Pinnaces, like fighter planes today, were a sign of war.

On their second day at Port Pheasant, another privateer, Captain Ranse of the Isle of Wight, appeared with two captured Spanish ships. Drake and Ranse promptly decided to combine forces for a daring attack which could make the fortune of every Englishman there—an attack upon Nombre de Dios.

Here, awaiting shipment to Spain, was stored the gold and silver carried from Peru to Panama and thence overland to Nombre de Dios. It was a small but important town founded in 1510 by weary Spanish explorers who, stumbling up out of marshes to slightly higher ground, exclaimed "Nombre de Dios ('in the Name of God!') let us settle here."

Drake could hope to approach Nombre de Dios undetected because the Spaniards were trying to hold a huge empire in the New World very thinly, with few men. Much tropical wilderness lay between their fortified posts.

On July 20, 1572, Drake and Ranse sailed west toward Nombre de Dios. On their way they seized two small Spanish ships and their crews of seven Negro slaves. From these slaves Drake may have had his first news of the Cimaroons.

The Cimaroons were escaped black slaves who hated their former masters, the Spaniards. They had married Indian women, established two independent kingdoms along the isolated mountain road from Panama to Nombre de Dios, and attacked Nombre de Dios itself. (*Cimarones* is Spanish for "Hill-Men.") An alliance between English pirates and Cimaroons was enough to make any peace-loving Spaniard turn pale.

"Willing to use those Negroes well," Drake landed his seven captives on the shore so they could flee to join the Cimaroons. Then he chose twenty of Ranse's men and fifty-three of his own—including John Drake and the daredevil Oxenham—for the assault. He embarked them in pinnaces and navigated to an island only half a day's voyage from Nombre de Dios.

"Remember the greatness of the hope of good things [in Nombre de Dios] . . . the weakness of the town . . . and the hope . . . to recompense . . . wrongs!" Drake exhorted his men on the morning of July 28, 1572. The youths eagerly drilled with firepikes, muskets, pikes, and bows and arrows that were distributed to them. (Firepikes were pikes around which tow was wrapped, to be lighted to help the attack and confuse the enemy.) In the afternoon, they set sail for Nombre de Dios.

How these seventy-four Englishmen surprised several hundred Spaniards in Nombre de Dios is related in the following selection from Philip Nichols's *Sir Francis Drake Revived* (1625).

That afternoon, he [Drake] causeth us to set sail for Nombre de Dios, so that before sunset we were as far as Rio Francisco. Thence, he led us hard aboard the shore, that we might not be described of the Watch House, until that being come within two leagues of the point of the bay, he caused us to . . . cast our grappers [grappling irons], riding so until it was dark night.

Then we weighed again, and set sail, rowing hard aboard the shore . . . till we recovered the point of

Nombre de Dios, a town on the east coast of Panama. From here, gold and silver from Peru were shipped by galleon to Spain. Drake raided Nombre de Dios in 1572.

the harbour under the high land. There, we stayed all silent; purposing to attempt the town in the dawning of the day. . . .

But our Captain . . . finding that our people were talking of the greatness of the town, and what their strength might be . . . thought it best to put these conceits out of their heads, and therefore to take the opportunity of the rising of the moon that night, persuading them that "it was the day dawning." By this occasion . . . we arrived [at the town] by three of the clock after midnight.

At what time it fortuned that a ship of Spain . . . which had but lately come into the bay; and had not yet furled her sprit-sail (espying our four pinnaces . . .) sent away her gundeloe [gondola] towards the town, to give warning. But our Captain perceiving it, cut betwixt her and the town, forcing her to go to the other side of the bay; whereby we landed without impeachment [hindrance], although we found one gunner upon the Platform [battery] in the very place where we landed. . . .

There we found six great pieces of brass ordnance [cannon]. . . . We presently dismounted them. The gunner fled. The town took alarm . . . as we perceived, not only by the noise and cries of the people, but by the bell ringing out, and drums running up and down the town.

Our Captain . . . left twelve to keep the pinnaces; that we might be sure of a safe retreat, if the worst befell. [After hastening to a hill to the east of the town to make sure no cannon had been planted there, Drake returned and] appointed his brother, with JOHN OXNAM [Oxenham] and sixteen other of his men to . . . enter [Nombre de Dios] near the easter[n] end . . . : himself with the rest, would pass up the broad street into the Market Place, with sound of drum and trumpet.

The Firepikes . . . served no less for fright to the enemy than light of our men, who . . . might discern every place . . . as if it were near day: whereas the inhabitants stood amazed . . . imagining, by reason of our drums and trumpets sounding in so sundry places, that we had been a far greater number than we were.

Yet . . . the soldiers and inhabitants had put themselves in arms, and brought their companies . . . at the south-east end of the Market Place . . . not far from the gate of the town . . . leading towards Panama. . . . The soldiers and [inhabitants] presented us with a jolly hot volley of shot . . . levelling very low, so as their bullets ofttimes grazed on the sand.

We . . . having discharged our first volley of shot . . . feathered them with our arrows . . . fine roving shafts . . . [then] came to the push of pike. . . . In short time [we] took such order among these gallants (some using the butt-end of their pieces instead of other weapons) . . . [that] they casting down their weapons, fled all out of the town by the gate aforesaid. . . .

Our Captain having taken two or three Spaniards . . . commanded them to shew him the Governor's House, where . . . was the ordinary place of unlading the moiles [mules] of all the treasure which came from Panama . . . although . . . the gold, pearl, and jewels . . . was [afterward] carried to the King's Treasure House. . . .

At our coming to the Governor's House, we found the great door . . . even then opened, a candle lighted upon the top of the stairs. . . . By means of this light we saw a huge heap of silver in that nether [lower] room; being a pile of bars of silver of . . . seventy feet in length, of ten feet in brea[d]th, and twelve feet in height, piled up against the wall, each bar . . . between thirty-five and forty pounds in weight.

At sight hereof, our Captain commanded . . . that none of us should touch a bar of silver; but stand upon our weapons, because the town was full of people, and there was in the King's Treasure House near the water side, more gold and jewels than all our four pinnaces could carry.

Drake's youths had entered town shouting "Victory, victory for the Queen of England!" But now, just as they had been nervous before the attack, they became panicky at rumors that their pinnaces were about to be captured and that 150 Spanish soldiers were in Nombre de Dios. Musket fire continued from Spanish snipers in the houses.

To give his men something else to think about, Drake tried to speed up breaking into the King's Treasure House.

Presently . . . a mighty shower of rain, with a terrible storm of thunder and lightning, fell, which poured down so vehemently . . . that before we could recover the shelter of a . . . pent-house at the western end of the King's Treasure House . . . some of our bow-strings were wet, and some of our match and powder hurt! . . .

As soon as the storm began to assuage of his fury (which was a long half hour) willing to give his men no longer leisure to demur of those doubts . . . [our Captain] stept forward commanding his brother, with JOHN OXNAM and the company appointed them, to break the King's Treasure House. . . .

But as he [Drake] stepped forward, his strength and sight and speech failed him, and he began to faint for want of blood, which . . . had, in great quantity, issued upon the sand, out of a wound received in his leg in the first encounter . . . to the . . . dismay of all our company. . . .

And therefore . . . they . . . having given him somewhat to drink . . . and . . . bound his scarf about his leg . . . entreated him to . . . go with them aboard . . . to have his wound . . . dressed. . . .

This when they could not persuade him unto . . . they . . . with force mingled with fair entreaty . . . bare him aboard his pinnace, and so abandoned a most rich spoil . . . to preserve their Captain's life. . . . Thus we embarked by break of the day (29th July), having besides our Captain, many of our men wounded, though none slain but one Trumpeter.

THE GREAT SOUTH SEA

Drake had broken into the Spanish treasure house and caught a tantalizing glimpse of the gold and silver, but he had to leave empty-handed. Though the attack brought him great notoriety, it was actually a fiasco.

Some historians even doubt that there was much silver in Nombre de Dios at this time (July) because the Plate Fleet had sailed for Seville a few weeks before. Treasure usually accumulated through the year in Panama and was not taken across the Isthmus to Nombre de Dios until January or February.

But Drake was not discouraged. Since he had failed at the treasure house, he would try to seize a transport—that is, one of the "mule trains" which carried the gold and silver from Panama to Nombre de Dios. Roads were closed during the rainy season, however, from April to December, so Drake passed that time capturing Spanish merchant ships and threatening Spanish ports on the Caribbean coast of South America.

When he found he lacked enough men for both his ships and his pinnaces, Drake had his carpenter secretly bore holes in the smaller *Swan* so that it sank—to Drake's pretended astonishment. (Cortes had destroyed his ships by the same method on the coast of Mexico before marching into the interior to conquer Montezuma.)

During these months, Drake lost his two brothers. John Drake was mortally wounded in October, 1572, when he rashly attempted to capture an armed Spanish ship with a handful of unarmed followers. Shortly afterward, young Joseph died in an epidemic of yellow fever. Only forty-seven of Drake's original seventy-three men remained.

But Drake clung stoically to his project, and he secured the help of the Cimaroons. Slowly the epidemic passed, and the rains ceased. On February 3, 1573, Drake set out with eighteen Englishmen and thirty Cimaroons to cross the mountains and ambush a treasure train near Panama, where the Spaniards might be off guard.

The expedition wound upward along paths so heavily forested the men saw only dark green overhead. They passed the Cimaroon capital in the woods, and on February 11, 1573, reached a clearing on the highest ridge of the Isthmus.

The following selection, from Philip Nichols's *Sir Francis Drake Revived* (1625), describes the sight Drake saw here and would never forget.

All the way was through woods very cool and pleasant, by reason of those goodly and high trees. . . . This [also] gave a special encouragement unto us all, that we understood there was a great Tree about the midway, from which, we might at once discern the North Sea [the Atlantic] from whence we came, and the South Sea [the Pacific] whither we were going.

The fourth day following (11th February) we came to the height of . . . a very high hill, lying

East and West, like a ridge between the two seas, about ten of the clock: where [PEDRO] the chiefest of these Cimaroons took our Captain by the hand, and prayed him to follow him, if he was desirous to see at once the two seas. . . .

Here was that goodly and great high Tree, in which they had cut and made divers steps, to ascend up near unto the top, where they had also made a convenient bower, wherein ten or twelve men might easily sit: and from thence we might, without any difficulty, plainly see the Atlantic Ocean whence now we came, and the South Atlantic [the Pacific Ocean] so much desired. South and north of this Tree, they had felled certain trees, that the prospect might be the clearer; and near about the Tree there were divers strong houses that had been built long before . . . by . . . Cimaroons . . . which usually pass that way. . . .

After our Captain had ascended to this bower, with the chief Cimaroon, and having . . . by reason of the brize [breeze], a very fair day, had seen that sea, of which he had heard such golden reports: he "besought Almighty GOD of His goodness, to give him life and leave to sail once in an English ship, in that sea!"

And then calling up all the rest of our [17 English]men, he acquainted JOHN OXNAM especially with this his petition and purpose. . . . [Oxenham] presently [at once] protested, that "unless our Captain did beat him from his company, he would follow him, by GOD'S grace!"

Thus all, thoroughly satisfied with the sight of the seas, descended; and after our repast, continued our ordinary march through woods.

Since geographical knowledge reached England late and the Spanish and Portuguese kept many discoveries secret, it is possible that Drake was not sure of the extent of the "South Sea" and that, in a sense, he rediscovered the Pacific. His resolution to "sail . . . in that sea!" links this Panama adventure to his later voyage around the world.

Of course the Pacific had been discovered in 1513 by Balboa, standing on this same central cordillera—or, possibly, a year earlier by Antonio de Abreu, a Portuguese who may have voyaged from Indonesia into its westernmost reaches. The Spaniards called it the South Sea because it lies south of Panama (the Isthmus runs east and west).

Until Magellan's voyage in 1520, however, men thought that the Pacific was much narrower than it really is, over twice the width of the Atlantic. Magellan thought he could cross it to Indonesia in two or three weeks—but the crossing took him three months! He renamed the ocean the Pacific because he found its waters so calm when he entered them.

ROBERT PIKE SPOILS ALL

Two days later Drake led his men into a dangerous open country of pampas. Then for two days more they made their way stealthily through the high grass until they saw, far off, a rocky peninsula covered with white houses, a harbor, and ships flying the red and yellow standards of Castile—the port of Panama.

Drake drew a deep breath. That afternoon he sent a Cimaroon into town to spy, and the Cimaroon returned with electrifying news. A mule train would depart that very night; in its vanguard would be the treasurer of Lima, his daughter, and nine mules laden with gold and jewels!

Drake first called upon "Almighty God" and then marched his men back several miles to form an ambush. On the way they captured a Spanish sentinel who was asleep.

The night attack on the mule train would be a *camisada* (from Spanish *camisa*—"shirt"). That is, Drake ordered his men to put their white undershirts on over their other clothes so that they could see each other, like Halloween ghosts, in the dark. He posted them on either side of the dirt road, then crouched by them, his hand on the whistle that hung around his neck. Everyone strained his ears for the first distant tinkle of bells on the mule train.

The following selection, from Philip Nichols's *Sir Francis Drake Revived* (1625), tells what happened next.

Being at the place appointed, our Captain with half his men [8 English and 15 Cimaroons], lay on one side of the way, about fifty paces off in the long grass; JOHN OXNAM with the Captain of the Cimaroons, and the other half, lay on the other side . . . but so far behind, that . . . the former company [Drake's] might take the foremost mules by the heads, and [Oxenham's] the hindmost [mules] . . . and especially that if we should have . . . to use our weapons . . . we might be sure not to endamage our fellows.

We had not lain thus in ambush much above an hour, but we heard the *Recuas* [mule trains] coming both from the city [of Panama] to Venta Cruz, and from Venta Cruz to the city [of Panama], which hath a . . . great trade, when the fleets are there. We heard them by . . . [their] deep-sounding bells, which in a still night, are heard very far off.

Now though there were . . . great charge given . . . that none of our men should shew or stir themselves, but let all that came from Venta Cruz to pass quietly . . . because . . . they brought nothing but merchandise from thence: yet one of our men, called ROBERT PIKE, having drunken too much aqua vitae [liquor] . . . forgot himself, and enticing a Cimaroon forth with him was gone hard to the way [far ahead] with intent to have shown his forwardness [boldness]. . . .

And when a cavalier from Venta Cruz, well mounted, with his page running at his stirrup, passed by . . . he [Robert Pike] rose up to see what he was: but the Cimaroon of better discretion pulled him [Robert Pike] down and lay upon him. . . .

Yet by this, the gentleman had taken notice by seeing one half all in white: for that we had all put our shirts over our other apparel, that we might be sure to know our own men in the pell mell in the night. . . . The cavalier, putting spurs to his horse, rode [off] . . . to give advertisement to others. . . .

Our Captain who had heard and observed by reason of the hardness of the ground and stillness of the night, the change of this gentleman's trot to a gallop, suspected that he was discovered, but could not imagine by whose fault. . . . And therefore considering that it [the Spaniard's galloping off] might be, by reason of the danger of the place . . . we lay still in expectation of the Treasurer's coming.

And he had come forward to us, but that this horseman meeting him, and . . . making report to him, what he had seen . . . and what he conjectured to be most likely: viz., that the said Captain DRAKE . . . was . . . come by land, in covert through the woods, unto this place . . . and thereupon persuaded him [the treasurer] to turn his *Recua* [mule train] out of the way, and let the other *Recuas* which were coming after to pass on. They were . . . loaded but with victuals for the most part . . . and yet they should serve to discover [Drake's men]. . . .

The other two *Recuas* were no sooner come up to us, but being stayed and seized on. One of the Chief Carriers . . . told our Captain by what means we were discovered, and counselled us to shift for ourselves betimes, unless we were able to encounter the whole force of the city and country. . . .

It pleased us but little, that we were defeated of [failed to capture] our golden *Recua*, and that in these we could find not past some two horse-loads of silver; but it grieved our Captain much more, that he was discovered, and that by one of his own men.

Drake was now in great danger. Spanish soldiers would soon appear from Panama, in the south, while to the north his retreat was blocked by the fortified town of Venta Cruz. Drake decided quickly—on "the shortest and readiest way." He would storm Venta Cruz, and if he got through, he could continue northward to his Caribbean base.

The white-shirted English and Cimaroons passed silently along the narrow road, between woods "as thick as our thickest hedges in England," until a Spanish sentinel challenged "*Qué gente?*"

"Englishmen!" shouted Drake and fired his pistol at him.

After a sharp skirmish, the Spaniards fled and Drake led his men into Venta Cruz. Three Spanish gentlewomen there were in a panic until Drake personally assured them they would not be harmed.

Drake's men held Venta Cruz for an hour and a half. Then they melted into the forest to the north, and the Spaniards did not follow.

RENDEZVOUS AT RIO FRANCISCO

Drake was finding gold and silver on the Spanish Main to be almost as elusive as the gold and silver at the end of the rainbow. But he would not admit defeat—and as long as he stayed, the Spaniards "even in their beds lay in great and continual fear of our Captain," as Drake was told.

On March 25, 1573, while Drake cruised in the Caribbean on the lookout for more prizes, he met a French privateer in distress. Its men were ill from lack of water, and its Huguenot (French Protestant) captain was depressed by news of the massacre of thousands of his fellow religionists by French Catholics on St. Bartholomew's Eve (August 23, 1572).

Drake offered Captain Tetû water, sympathy, and a

French, Dutch, and English traders on the Spanish Main discovered they could raid sparsely protected Spanish towns of their gold and silver.

plan to get rich quick. He asked the melancholy Frenchman to help him ambush a silver mule train near Nombre de Dios. Perhaps the guards would be careless so close to the end of their journey from Panama.

Captain Tetû agreed. He would make his seventy men available and be content with half the profits. Drake's youths numbered only thirty-one, but they were battle-hardened now, and their morale remained high. Drake's words of encouragement and his understanding of their needs, as one who had been a common seaman himself, nerved them for a final strike against the huge but thinly-defended empire.

The following selection, from Philip Nichols's *Sir Francis Drake Revived* (1625), describes Drake's third surprise attack in Panama.

As soon as we could furnish ourselves and refresh the French company . . . taking twenty of the French and fifteen of ours with our Cimaroons, leaving both our ships in safe road, we manned our frigate and two pinnaces . . . and went towards Rio Francisco. . . .

[There] both Captains landed (31st March) with such force as aforesaid [*i.e.*, twenty French, fifteen English, and the Cimaroons], and charged them that had the charge of the pinnaces to be there the fourth day next following without any fail. And thus knowing that the carriages [mule trains] went now daily from Panama to Nombre de Dios; we proceeded in covert through the woods, towards the highway that leadeth between [Panama and Nombre de Dios]. . . .

When we had come within an English mile of the [high]way, we stayed all night, refreshing ourselves, in great stillness, in a most convenient place: where we heard the carpenters, being many in number, working upon their ships, as they usually do by reason of the great heat of the day in Nombre de Dios; and might hear the mules coming from Panama. . . .

The next morning (1st April), upon hearing of that number of bells, the Cimaroons rejoiced exceedingly. . . . Now they all assured us, "We should have more gold and silver than all of us could bear away." . . .

For there came three *Recuas* [mule trains], one of 50 mules, the other two, of 70 each, every [one] of which carried 300 lbs. weight of silver; which in all amounted to near thirty tons [*i.e.*, 190 mules, with 300 pounds each—about 57,000 pounds of silver].

We putting ourselves in readiness, went down near the [high]way to hear the bells; where we . . . took such hold on the heads of the foremost and hindmost mules, that all the rest stayed and lay down, as their manner is.

These three *Recuas* were guarded with forty-five soldiers . . . fifteen to each *Recua*, which caused some exchange of bullets and arrows for a time; in which conflict the French Captain was sore wounded with hail-shot in the belly, and one Cimaroon was slain: but in the end, these soldiers thought it the best way to leave their mules with us, and to seek for more help aboard.

In which meantime we took some pain to ease some of the mules which were heaviest laden. . . . And because we ourselves were somewhat weary, we were contented with a few bars and quoits of gold, as we could well carry: burying about fifteen tons of silver, partly in burrows which the great land crabs had made in the earth, and partly under old trees which were fallen thereabout, and partly in the sand and gravel of a river, not very deep of water. . . .

The writer employs a kind of joking understatement. The "few bars and quoits of gold" were worth close to $400,000.

When . . . we had spent some two hours . . . and were ready to march back . . . we heard both horse and foot coming as it seemed to the mules: for they [the returning Spaniards] never followed us, after we were once entered the woods, where the French Captain, by reason of his wound, not able to travel farther, stayed, in hope that some rest would recover him better strength.

But after we had marched some two leagues, upon the French soldiers' complaint, that they missed one of their men also . . . it was found that he had drunk much wine, and . . . had lost himself in the woods. And as we afterwards knew, he was taken by the Spaniards that evening; and upon torture, discovered unto them where we had hidden our treasure.

We continued our march all that and the next day (2nd and 3rd April) towards Rio Francisco, in hope to meet with our pinnaces; but when we came thither, looking out to sea, we saw seven Spanish pinnaces, which had been searching all the coast thereabouts; whereupon we mightily suspected that they had taken . . . our pinnaces. . . .

But the night before, there had fallen very much rain, with much westerly wind, which as it enforced the Spaniards to return home the sooner . . . so it kept our pinnaces, that they could not keep the appointment; because the wind was contrary. . . .

Our Captain seeing the [Spanish] shallops, feared least having taken our pinnaces, they had compelled our men by torture to confess where his frigate and ships were. Therefore in this distress and perplexity, the company misdoubting that all means of return to their country were cut off . . . our Captain comforted and encouraged us all, saying,

"We should venture no farther than he did. It was no time now to fear . . . ! If the enemy have prevailed against our pinnaces, which GOD forbid! yet they must have time to search them, time to examine the mariners, time to execute their resolution after it is determined. Before all these times be taken, we may get to our ships, if ye will! though not possibly by land, because of the hills, thickets, and rivers, yet by water. Let us, therefore, make a raft with the trees that are here in readiness . . . being brought down the river . . . this last storm, and put ourselves to sea! I will be one, who will be the other?"

JOHN SMITH offered himself, and two Frenchmen that could swim very well . . . the Cimaroons likewise. . . . The raft was fitted and fast bound; a sail of a biscuit sack prepared; an oar was shaped out of a young tree to serve instead of a rudder, to direct their course before the wind.

At his departure, he comforted the company, by promising, that "If it pleased GOD, he should put his foot in safety aboard his frigate, he would, GOD willing, by one means or other get them all aboard, in despite of all the Spaniards in the Indies!"

In this manner pulling off to the sea, he sailed some three leagues, sitting up to the waist continually in water, and at every surge of the wave to the arm-pits, for the space of six hours, upon this raft: what with the parching of the sun and what with the beating of the salt water, they had all of them their skins much fretted away.

At length GOD gave them the sight of two pinnaces turning towards them with much wind . . . and [our Captain] did cheerfully declare to those three with him, that "they were our pinnaces! and that all was safe, so that there was no cause of fear!"

But . . . the pinnaces not seeing this raft . . .

by reason of the wind and night growing on, were forced to run into a cove behind the point . . . for that night: which our Captain seeing, and gathering . . . that they would anchor there, put his raft ashore, and ran by land about the point, where he found them; who, upon sight of him, made as much haste as they could to take him and his company aboard.

For our Captain (of purpose to try what haste they could and would make in extremity), himself ran in great haste, and so willed the other three with him; as if they had been chased by the enemy. . . .

And after his coming aboard, when they demanding "How all his company did?" he answered coldly, "Well!" They all doubted [feared] that all went scarce well. But he willing to rid all doubts, and fill them with joy, took out of his bosom a quoit of gold, thanking GOD that "our voyage was made!"

And to the Frenchmen he declared how their Captain indeed was left behind, sore wounded and two of his company with him: but it should be no hindrance to them.

That night (4th April) our Captain . . . rowed to Rio Francisco: where he took the rest in, and the treasure which we had brought with us: making such expedition, that by dawning of the day, we set sail back again to our frigate, and from thence directly to our ships. [There], as soon as we arrived, our Captain divided by weight, the gold and silver into two even portions, between the French and the English.

PLYMOUTH CHURCH IS EMPTIED

When some of Drake's men went back to Rio Francisco for the buried silver, they found the ground all torn up and the treasure gone. They also learned of the capture of Captain Tetû, but not of his death.

But the voyage was "made." The English share of the gold pesos and silver bars taken from the mule train amounted to nearly $200,000. And Drake had captured two swift new Spanish frigates in which to make his getaway to Plymouth.

He had also done more than simply enrich himself and his men. Although no longer as populous and prosperous as it had been, the Spanish Main was still the outer court of the world's most powerful empire. That Drake and his handful of seamen had been able to defy the guardians of this court, to disrupt coastal trade, and to make raids into the interior of Panama suggested that Spain might not be invincible.

If one day Drake should extend his daring to the Pacific, as he had vowed he would, the shock to Spanish and the encouragement to English morale might affect the future of the New World.

The following selection, from Philip Nichols's *Sir Francis Drake Revived* (1625), describes Drake's parting with the Cimaroons, his return voyage, and the sensation caused by his arrival in Plymouth—an arrival highly embarrassing to Queen Elizabeth.

About a day or two before our departure, our Captain willed PEDRO and three of the chiefest of the Cimaroons to go through both his frigates, to see what they liked; promising to give it them. . . .

And for their wives he would himself seek out some silks or linen . . . which while he was choosing out of his trunks, the scimitar which CAPTAIN TETÛ [the French Captain] had given to our Captain, chanced to be taken forth in PEDRO'S sight: which he seeing grew so much in liking thereof, that he . . . preferred it before all that could be given him.

Our Captain . . . desirous to content him, that had deserved so well . . . gave it him . . . who received it, with no little joy . . . and . . . desired our Captain to accept . . . four pieces of gold [in recompense]. . . . Thus with good love and liking we took our leave of that people. . . .

[At] Cape St. Antonio [Cuba] . . . we refreshed ourselves, and beside great store of turtle eggs, found by day in the [sand], we took 250 turtles by night. We powdered [salted] and dried some of them, which did us good service. . . .

There were, at this time, belonging to Cartagena, Nombre de Dios, Rio Grande, Santa Marta, Rio de la Hacha, Venta Cruz, Veragua, Nicaragua, the Honduras, Jamaica, &C., above 200 frigates . . . which all had intercourse between Cartagena and Nombre de Dios. The most of which, during our abode in those parts, we took; and some of them, twice or thrice each: yet never burnt nor sunk any, unless they were made out Men-of-war against us, or . . . [used] to entrap us.

And of all the men taken in these several vessels, we never offered any kind of violence to any, after they were once come under our power; but either presently dismissed them in safety, or keeping them with us some longer time . . . at last, the danger of their discovering where our ships lay being over past . . . we set them also free. . . .

From this Cape of St. Antonio . . . [we sailed] the directest and speediest way homeward. . . . For whereas our Captain had purposed to touch at Newfoundland, and there to have watered . . . yet GOD Almighty so provided for us, by giving us good store of rain water, that we were sufficiently furnished: and, within twenty-three days, we passed from the Cape of Florida, to the Isles of Scilly, and so arrived at Plymouth, on Sunday, about sermon time, August the 9th, 1573.

At what time, the news of our Captain's return . . . did so speedily pass over all the church, and surpass their minds with desire and delight to see him, that very few or none remained with the Preacher. All hastening to see the evidence of GOD'S love and blessing towards our Gracious Queen and country, by the fruit of our Captain's labour and success.

Chapter Two

A COMMISSION FROM THE QUEEN

I would gladly be revenged
on the King of Spain,
for divers injuries
that I have received.
—QUEEN ELIZABETH I

Two

THE STRAIT OF MAGELLAN

When Drake returned from his Panama expedition, he found that the winds of diplomacy had again shifted. English mariners still distrusted Spain, but Elizabeth and Philip II had decided to try to settle their differences and improve relations between the countries. Drake's arrival at this moment with his pockets full of plundered Spanish pesos was very awkward for the Queen.

Probably on a hint from some court official, Drake disappeared—for two years. (Neither he nor the Queen wanted to have to give the Spanish gold back as part of the negotiated settlement.) He went into hiding, probably in a little creek later called Drake's Pool, in Queenstown Harbor, in southern Ireland.

"Our Captain [Drake] was glad to come into Ireland for fear of . . . the Council because of his Indian voyages," said a contemporary.

For a short time, lawful trade between England and Spain revived and prospered. But even during this period, projects which would trespass upon Spain's New World empire were discussed. Merchants still sought new markets overseas because of the collapse of their cloth trade with Antwerp. Noblemen and naval leaders, members of the Elizabethan "Establishment" from the Queen down, were still fascinated by gold in the Americas and by spices in the East.

There were two main schemes for acquiring this wealth. One was a voyage to the Orient by the as yet undiscovered Northwest Passage. The other was a voyage by the Southwest Passage, that is, through the Strait of Magellan, with a few extra dividends from trade or piracy en route.

At first the Northwest Passage was more favored to avoid a direct confrontation with Spain. Sir Humphrey Gilbert had been pushing this project since the mid 1560's, and in 1576, Martin Frobisher finally sailed on the first of three well-organized expeditions to the Hudson Bay area. After his first voyage, Frobisher was certain that he had found the passage just north of Hudson Strait. But he also brought back ore rumored to contain gold, which distracted him from exploring farther west.

Yet as early as 1574, the Southwest Passage also had its champions. In that year, Richard Grenville, William Hawkins, and "divers gentlemen of the west" proposed a voyage through the Strait of Magellan for trade and settlement in lower South America. Possibly Grenville also wished to explore Terra Australis, the southern continent thought to begin with Tierra del Fuego; more likely—although this was carefully left unsaid—he planned to capture some Spanish treasure ships in the Pacific.

Queen Elizabeth was quite capable of grasping the unspoken intentions of her "hawks." She first gave a license for the voyage, then withdrew it—apparently deciding she did not wish to endanger the fragile détente with Philip II. But Grenville's proposal showed that the southwestern venture was in the air. It had the support of the leading English geographer, John Dee, and would soon arouse the interest of Francis Drake.

In 1576, when relations between England and Spain

were again worsening, Drake's old friend John Oxenham was able to slip away for another raid in the Caribbean. Oxenham managed to cross the Isthmus, build a pinnace, and seize prizes in the Pacific. On his return to the Isthmus he was captured by the Spaniards, who interrogated him closely about English plans for a southwestern voyage.

The following selection, from "Declaration of John Oxenham before the Inquisition at Lima" (Nuttall, Zelia, ed., *New Light on Drake*), gives information about early plans for a voyage to the Strait of Magellan and into the Pacific—a voyage similar to the one Drake would later make.

Questioned whether, while in England or since he [John Oxenham] had left there he had heard or understood that Queen Elizabeth or any other person had entertained the project to arm a certain number of vessels for the purpose of establishing settlements, or for other purposes, on the coast of the North Sea [the Atlantic], or in the region of the Strait of Magellan or on the coast of the South Sea [the Pacific], he [Oxenham] answered that four years ago an English knight named Richard Grenville, who lives at a distance of a league and a half from Plymouth, and is very rich, applied to the Queen for a license to come to the Strait of Magellan and to pass to the South Sea [the Pacific], in order to search for land or some islands where to found settlements, because, in England, there are many inhabitants and but little land.

The Queen gave him the license and witness saw it.

It was very large. The said Grenville bought two ships, and was about to buy two or three more, when the Queen revoked the license, because she had learnt that beyond the Strait of Magellan there were settlements made by Spaniards, who might do them harm.

The said Grenville sold the ships, after the license had been taken from him. Previously to that, he had spoken many times with witness [Oxenham], trying to persuade him to accompany him, but witness did not wish to do so. Grenville's project was to come and found a settlement on the River Plate [Rio de la Plata, Argentina] and then pass the Strait and establish settlements wherever a good country for such could be found.

Witness [Oxenham] thinks that if the Queen were to give a license to Captain Francis Drake he would certainly come and pass through the Strait, because he is a very good mariner and pilot, and there is no better one than he in England who could accomplish this. Witness thinks that the Queen will not, as long as she lives, grant the license, but that, after the Queen's death, there will certainly be someone who will come to the Strait. The said Captain Francis had often spoken to witness saying that if the Queen would grant him the license he would pass through the Strait of Magellan and found settlements over here in some good country.

Questioned with how many ships it would be possible for Francis Drake to come to the Strait he answered that with the aid of his relatives and companions he might be able to bring two or three vessels but that, after discovering a good country, they would be able to come with more ships. Witness said that Captain Francis discussed this subject with him.

Questioned whether they had discussed how, and by what route, they were to return to England after having passed through the Strait, he said that it seemed to him that some said that it was to be by the same Strait, but others said that there was a route through another Strait that passed into the North Sea [the Atlantic], but nobody knows this for a certainty or has passed through it.

A DRAFT PLAN

Drake reappeared in 1575. In April of that year he is listed as captain of a ship serving under the Earl of Essex, who had a commission to subdue and plunder the Irish. On this expedition, Drake met a gentleman-adventurer, Thomas Doughty, who could help him achieve his goal of sailing on the South Sea.

Thomas Doughty was described by a contemporary as "a sweet orator, a pregnant [learned] philosopher, a good gift for the Greek tongue . . . a sufficient secretary to a noble personage . . . an approved soldier"—in other words, a complete Elizabethan gentleman. Drake would have the financial backing of wealthy Plymouth shipowners, but Doughty could add influence at court—the interest of powerful men like the Queen's new favorite, Sir Christopher Hatton, whose secretary he became.

In the fall of 1575, Drake was in London, seeking support for a southwestern voyage. He later said that a letter from the Earl of Essex opened important doors to him, but Doughty claimed that his employer, Hatton, was responsible for the breakthrough. In any event, by the spring of 1577, Queen Elizabeth was once more angry enough with Philip—because the new Spanish governor in the Netherlands was playing the same game of conspiracy as the old ambassador—to approve a tentative plan for the project.

By great good fortune a badly burnt copy of this draft plan was discovered in 1930. It outlines a voyage to lower South America through the Strait of Magellan and halfway up the west coast of Chile for purposes of trade and exploration—possibly to find uninhabited areas for future colonization. Some scholars think the plan indicates exploration not of Chile but of Terra Australis. It may later have been expanded to include a journey across the Pacific to the Spice Islands and a search for the western entrance to the Northwest Passage.

The following selection, from Kenneth R. Andrews's *Drake's Voyages*, is the fragment of the draft plan which describes the route and purposes of the voyage. Letters and words supplied by conjecture are printed in italics.

. . . *shall enter the Strait of Magell*anas *lying in* 52 *degrees of* the pole, and *having passed therefrom into* the South Sea then *he is to sail so* far to the northwards as *xxx* [30] *degrees seeking* along the said coast af*orenamed like* as of the other to find out pl*aces meet* to have traffic for the vent*ing of commodities* of these her Majesty's realms.

Wher*eas at present* they are not under the obedience of *any christian* prince, so there is great hope of *gold, silver,* spices, drugs, cochineal, and *divers other* special commodities, such as may *enrich her* Highness' dominions, and also *put* shipping a-work greatly.

And *having* gotten up as afore said in the xxx *degrees* in the South Sea (if it shall be thought *meet* by the afore named Francis Drake to pro*ceed so* far), then he is to return by the same way homewards as he went out. Which voyaging by God's favour is to be performed in xiii months, all though he should spend v months in tarrying upon the coasts, to get knowledge of the princes and countries there.

SECRET MEETING WITH ELIZABETH

The draft plan contains a sentence providing for secrecy: "The Queen's Majesty may be made privy to the truth of the voyage, and yet the colour [pretense] to be given out [of a voyage] for Alexandria." But there is nothing in writing about seizing Spanish ships.

There did not have to be. Drake understood his Queen, and the Queen understood Drake. She had canceled Grenville's license for a similar voyage because she suspected his piratical intentions; yet now she approved the same venture for Drake, a more notorious sea raider.

Knowing that she was playing a dangerous game, Elizabeth insisted the project be hidden both from Spaniards and from cautious Lord Burleigh, champion of a peace policy. If Drake were caught, he would be in grave danger; his government might disown him.

Later, at a crucial moment on the voyage when he thought he was faced with mutiny, Drake revealed to his men a secret interview he had had with the Queen. In this meeting, according to Drake, Elizabeth spelled out a major purpose of the expedition: she "would gladly be revenged on the King of Spain."

The following selection gives Drake's account of how he won approval for his voyage and of the famous interview with Elizabeth. It is from John Cooke's "Narrative" (Vaux, W. S. W., ed., *The World Encompassed by Sir Francis Drake*). Cooke, of whom nothing else is known, was a member of the expedition.

Indeed thus it was: My lord of Essex wrote in my commendations unto secretary Walsingham. . . . He [Essex] thought me in his letters to be a fit man to serve against the Spaniards, for my practice and experience that I had in that trade.

Whereupon indeed secretary Walsingham did come to confer with [me], and declared unto [me] that for that her Majesty had received divers injuries of the King of Spain, for the which she desired to have some revenge; and withal he showed me a plot [map] . . . willing me to set my hand, and to note down where I thought he might most be annoyed.

But I told him some part of my mind, but refused to set my hand to any thing, affirming that her Majesty was mortal, and that if it should please God to take her Majesty away, it might be that some prince might reign that might be in league with the King of Spain, and then will mine own hand be a witness against myself.

Map showing the route of Drake's trip around the world, made in 1590 by Hondius, a famous Flemish cartographer. At the top center of the map is a picture of Drake; below, a picture of the Golden Hind.

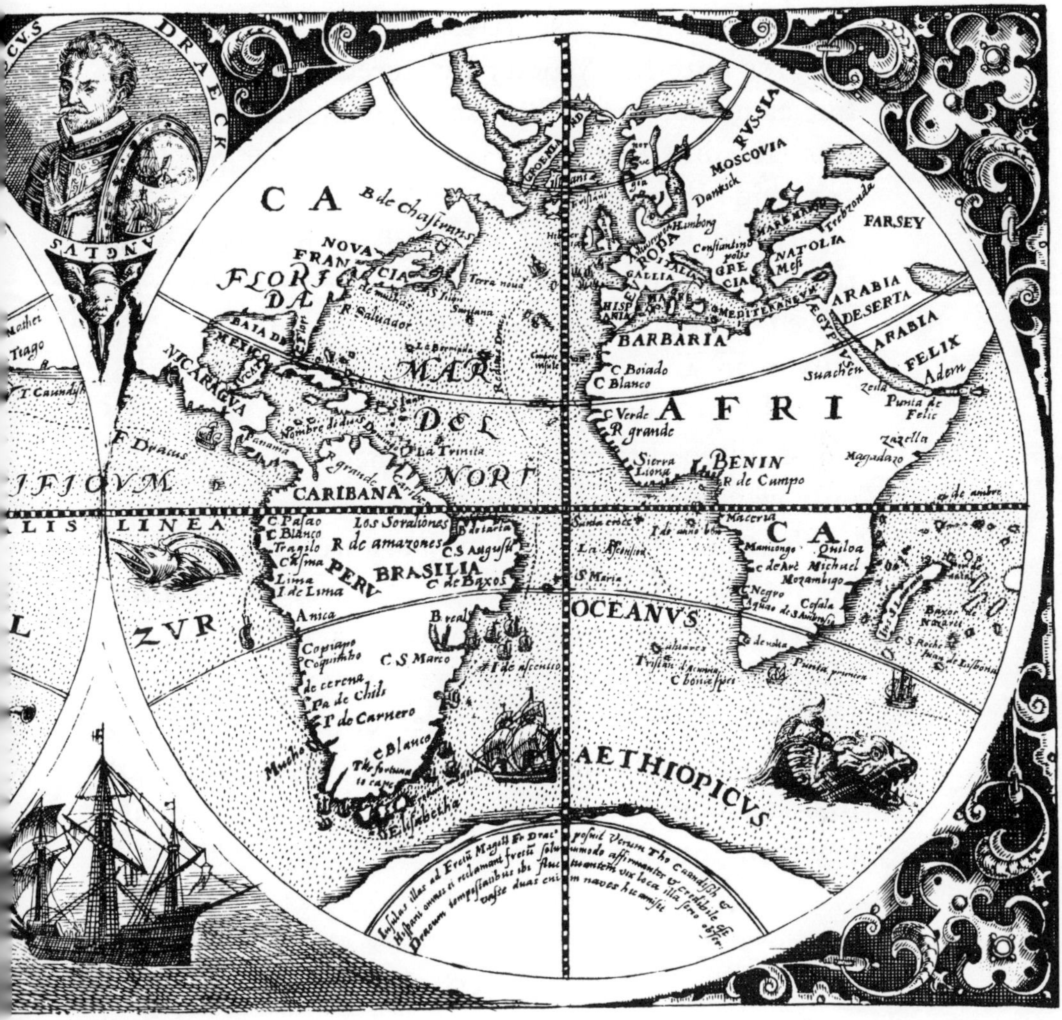
DRAECK
ANGLUS
CA
AFRI
CA
NOVA FRANCIA
FLORIDA
BAIA DE MEXICO
NICARAGVA
MAR DEL NORT
CARIBANA
BRASILIA
PERV
OCEANVS
AETHIOPICVS
ZVR
LINEA
RVSSIA
MOSCOVIA
FARSEY
NATOLIA
BARBARIA
ARABIA DESERTA
ARABIA FELIX
BENIN
GALLIA
C Blanco
C Verde
R grande
Sierra Liona
R de amazones
C S Augustin
C de Baxos
C S Marco
Pa de Chili
Pto de Carnero
C Blanco
Elisabetha
C bone spei
Aden
Dankick
Hamborg
Trebzonda
C Boiado
F Dracus
Los Soraliones
Lima
I de Lima
Anica
R de Cumpo
Quiloa
Mozambigo
Cofala
Punta de Felic
Magadazo

Then was I very shortly after and in an evening sent for unto her Majesty by secretary Walsingham, but came not to her Majesty that night, for that it was late; but the next day coming to her presence, [she said] these or the like words:

"Drake, so it is that I would gladly be revenged on the King of Spain, for divers injuries that I have received"—and [she] said farther that [I] was the only man that might do this exploit, and withal craved [my] advice therein. [I] told her Majesty of the small good that was to be done in Spain, but the only way was to annoy him by his Indies. . . .

Her Majesty did give [me] toward [my] charges . . . a bill of her Majesty's adventure of a 1000 crowns. . . . [I was also given] a bill of Master Hatton's adventure, and divers letters of credit. . . . Her Majesty did swear by her crown that if any within her realm did give the King of Spain hereof to understand (as she suspected but two) they should lose their heads therefore.

SHIPS AND ARMS

Drake and his financial backers, among whom was Queen Elizabeth, now fitted out a fleet that was very well equipped for both trade and plundering. Three ships were readied in Plymouth and two at London, where Thomas Doughty was seen conferring with Lord Burleigh.

All the ships were floating arsenals. The two largest—Drake's flagship, the *Pelican* (100 tons), and Captain John

A portrait of Queen Elizabeth I by an unknown artist.

Winter's *Elizabeth* (80 tons)—carried eighteen and sixteen cannon respectively. The *Marigold* (30 tons), *Swan* (50 tons), and *Benedict* (15 tons) also had cannon.

On Drake's flagship, a Spaniard later captured off the coast of Peru saw "fire-bombs and darts with a certain kind of artifice for setting fire to the sails of ships; chain balls for breaking topmasts and other deadly work; tackle and rigging and many arquebuses, corselets, pocket pistols, trappings, pikes and a great quantity of many different kinds of arms."

But the ships also carried goods for peaceful trade—knives, linen, woolen cloth, pickaxes, perfumed waters (for Eastern rulers), and "trifles" (probably including mirrors, basins, scissors, etc.). Two merchants went along as supercargoes in charge of these goods, which suggests that the Spice Islands had been added to Drake's itinerary. Food supplies consisted of biscuit, beef, pork, codfish, butter, cheese, oatmeal, rice, raisins, and dried peas.

Drake's young seamen were mostly English, but there were a few Frenchmen, Danes, Flemings, Scots, and Biscayans. He had on board the ship a smith, a shoemaker, a cooper, a carpenter, an apothecary (druggist), a tailor, a surgeon, and some musicians who played after the reading of Psalms.

There were also twelve gentleman-adventurers—Elizabethan gallants voyaging for the experience who did not expect to do any work. Altogether, Drake enlisted between 140 and 164 men, all volunteers.

The selection below describes Drake's ships and crews. It is from *The World Encompassed* (1628), a kind of sequel to *Sir Francis Drake Revived* (1625). *The World Encompassed* is anonymous, but it is based largely on notes left by the chaplain of the expedition, Francis Fletcher.

The said Captain *Francis Drake* having in a former voyage, in the years 72 and 73 . . . had a sight . . . of the South Atlantic [the Pacific], and thereupon . . . conceiving a . . . desire of sailing on the same, in an English bottom; he so cherished thenceforward this his noble desire and resolution . . . that notwithstanding he was hindered for some years, partly by secret envy at home, and partly by public service for his Prince [ruler] and country abroad [in Ireland] . . . yet . . . [in] 1577, by gracious commission from his sovereign, and with the help of divers friends adventurers, he had fitted himself with five ships.

1. The *Pelican*, admiral [flagship], burthen 100 tons, Captain-general *Francis Drake.*
2. The *Elizabeth*, vice-admiral [vice-flagship], burthen 80 tons, Captain *John Winter.*
3. The *Marigold*, a bark of 30 tons, Captain *John Thomas.*
4. The *Swan*, a flyboat of 50 tons, Captain *John Chester.*
5. The [*Benedict*], a pinnace of 15 tons, Captain *Thomas Moone.*

These ships he manned with 164 able and sufficient men, and furnished them also with such plentiful provision of all things necessary, as so long and dangerous a voyage did seem to require; and amongst the rest, with certain pinnaces ready framed, but carried aboard in pieces, to be new set up in smoother water, when occasion served.

Neither had he omitted to make provision also for ornament and delight, carrying to this purpose with him expert musicians, rich furniture (all the vessels for his table, yea, many belonging even to the Cookroom being of pure silver), and divers shows of all sorts of curious workmanship, whereby the civility and magnificence of his native country might, amongst all nations whithersoever he should come, be the more admired.

DESTINATION UNKNOWN

On November 15, 1577, in the midst of a war scare—a Spanish agent, de Guaras, had been caught corresponding with Mary Stuart—Drake set sail. However, almost immediately a Channel storm battered the fleet and drove the five ships back to Plymouth. After repairs, Drake weighed anchor a second time, December 13, 1577.

There was some confusion as to where this expedition was headed. Before he was detected writing political love letters to Queen Mary, de Guaras had reported that Drake was going to Scotland to kidnap young Prince James. Other Spaniards believed a raid on the West Indies was in the making.

Drake's seamen were told they would take on a cargo of currants in Alexandria. The gentleman-adventurers must have supposed that Drake would attack Spanish shipping—they knew Drake, and they could see the big bronze guns and knocked-down pinnaces aboard. But they had also been briefed about trade, in South America and/or the Spice Islands. If, for personal reasons, some of them later fell out with Drake, they might then quarrel about his goals.

The fleet that Drake now led southward toward the Strait of Magellan might also be pointed toward mutiny.

Chapter Three

MUTINY AT SAN JULIAN

[At] the Port de las Islas [San Julian] . . .
[Drake] killed a very noble gentleman
on account of the suspicion . . .
that the said gentleman intended to rise against him.
—NUÑO DA SILVA

Three

THE CAPE VERDE ISLANDS

The fleet rolled in the Atlantic swell—five small ships setting out through the "Green Sea of Darkness" to defy an empire. Drake led them on a southwest course past Gibraltar, the mountains of Morocco, and the white Sahara dunes.

He made frequent stops for food and water, a practice which would help protect the crews from scurvy and dysentery. At an island off Morocco he set up a pinnace, which then followed the fleet like a greyhound on leash and helped capture several small craft, Portuguese or Spanish.

As they sailed far down the bulge of Africa there rose in the south, at night, the six stars of the beautiful constellation which points toward the south celestial pole, the Southern Cross. No trouble developed on this "shakedown cruise," although the gentlemen kept aloof from the seamen, and Thomas Doughty seemed to regard himself almost as a co-commander with Drake.

On January 26, 1578, through the haze, Drake sighted the first of the ten Cape Verde Islands, a V-shaped cluster three hundred miles west of Africa's Cape Verde. The "V" lies on its side, with its open end toward the New World. Now the northeast trade wind carried the fleet along the lower arm of the "V" (>). Off Santiago, in spite of cannon fire from shore batteries, Drake's pinnace captured a heavily-laden Portuguese caravel.

Wine, cloth, and a dark-complexioned, graying captain who wore his long beard and long black coat with dignity made an attractive prize—especially the captain, Nuño da Silva. Da Silva was a veteran pilot who had made the crossing to Brazil many times and may have voyaged along the coast of Chile and Peru.

The following selection describes the capture of da Silva and the volcano of the next island, cone-shaped Fogo (Portuguese for "Fire"). It is from the manuscript narrative of the chaplain, Francis Fletcher, the same narrative that was used by the compiler of *The World Encompassed* . . . but which contains additional details omitted in the printed work (Vaux, W. S. W., ed., *The World Encompassed*).

Coming to the southerly cape of this island [Santiago], we discovered near to the seaside the town of St. James, with a castle and blockhouse . . . well mounted with ordinance. For two ships of Portugal being lately come forth of the harbor, bound for Brazil in a merchant voyage, we sent our pinnace to command them to repair to our fleet.

And our pinnace recovering the one which was farther off at the sea than the other, the castle applied the pinnace with great shot to defend the ship that was next them till she might get into harbor, and [to] rescue the other . . . but we brought it [the other] away without any harm done to our pinnace or men, being a ship of Portugal laden with singular wines . . . and . . . with woolens and linen clothes, silks, and velvets, and many other good commodities . . . [so] that she was the life of our voyage . . . [which would otherwise have been endangered by] the shortness of our provisions. . . .

We [next] . . . passed near to the Island of Fogo . . . which is so named of the Portuguese because it burneth continually in a most strange manner, far exceeding Aetna in my opinion. . . . For in the north part thereof ariseth a great hill . . . reaching into the air . . . some six English miles . . . in form like a steeple's spire . . . out of the concavity whereof . . . ariseth, as out of a chimney . . . a . . . thick smoke . . . and . . . abundance of flames . . . the light thereof . . . so great, that in the extremest darkness of the night it seemeth as noon day. . . .

The flame being dispersed, there followeth . . . such infinite numbers of pumice-stones scattered abroad in the air . . . that falling down they cover the water, and are there to be taken up as sponges swimming upon the face of the seas. Last of all is cast out abundance of heavy, black, hard substances congealed as smith's cinders, which . . . fall down the side of the spire with a great noise to the lower part, where resting, they increase continually the outside of the hill. . . .

Now [near the island of Brava, most southwesterly of the Cape Verdes] the Portuguese of the [captured] ship having been . . . set freely at liberty . . . we reserved to our own service only one of that company, one da Silva, their pilot, a man well traveled both in Brazil and most parts of India . . . who, when he heard that our travel was into Mar del Sur, that is, the South Sea [the Pacific], was most willing to go with us.

It is very doubtful that da Silva was "most willing to go" with the English—he had no choice. Likewise, the statement that there was a "shortness of provisions" in Drake's fleet is dubious—the fleet was provisioned for a voyage of a year and a half. But probably Drake's sailors were thirsty for da Silva's wine.

Drake kept da Silva's ship, renamed the *Mary*, as well as the Portuguese pilot. He had replaced the *Benedict* with a captured Spanish fishing vessel and now sailed with six ships.

STRIFE ON SHIPBOARD

As green Brava fell astern, the little fleet was swallowed up in the wide ocean—and nearly doomed by a series of explosive quarrels that broke out almost immediately. Cramped quarters on the small ships, the sultry heat and thunderstorms of the doldrums, class resentments, and personality conflicts—all combined to poison the atmosphere.

Some scholars believe that Thomas Doughty, leader of the "gentlemen," was also leader of a gentleman-adventurers' peace party that wished to emphasize trade rather than raid. It is true that John Winter, captain of the *Elizabeth*, which later turned back to England from South

America, made a legal declaration that he had taken part in the capture of da Silva's ship because of his fear of Drake. But he was being charged with piracy by the Portuguese ambassador and may just have been trying to excuse himself.

Whether the reason for the gentlemen's aloofness from the seamen was class feeling or dislike of Drake's piracy, they did rally around Doughty. And Drake, whether he shared his deckhands' class prejudice—he had come up from the ranks himself—or whether he recognized a threat to his and Elizabeth's goals, singled out Doughty as the sputtering, disaffected man who might set off a mutiny.

The following selection is from Captain John Winter's "Declaration About Nuño da Silva's Ship" (Nuttall, Zelia, ed., *New Light on Drake*).

The said John Winter sayeth that upon the coast of Africa near unto certain Islands of Cape Verde the said Francis Drake by himself and others by his power and forceable commandment did take a certain Portuguese Ship wherein were certain Madeira wines and what other goods he knoweth not.

And after the taking of the said Ship he the said Drake did put the men aland in a Pinnace and carried away the wines with the Ship for the relief and maintenance of himself and company being bent upon a long voyage of two years as he said, and as it was then supposed.

The taking of which ship I protest was utterly contrary to my good will which I could not let [prevent] . . . for that I had no authority there, but such as pleased the said Drake, to give and take away from me. . . . And being in great fear of my life if I should have contraried him . . . he would have punished me by death, for that his words and threatenings many times tended thereunto. . . . And for that I was . . . enforced to content myself with silence.

The next selection, from John Cooke's "Narrative" (Vaux, W. S. W., ed., *The World Encompassed*), gives details of the conflict in the fleet.

Now Master Drake having here somewhat satisfied his eye with the view of these commodities [the captured Portuguese goods], he committed the custody . . . of this prize [the Portuguese ship] unto Master Thomas Doughty as his good and especial friend, praying him in any case to see good order kept. . . .

It thus chanced that Master Drake had a brother (not the wisest man in Christendom who he put into this said prize, as also divers others. This Thomas Drake, as one more greedy of prey than covetous of honesty or credit . . . did not only break open a chest, but did dive suddenly into the same, that Master Doughty knew not how to discharge himself against the General [Drake] but by revealing it unto him [Drake].

Yet first Master Doughty called Thomas Drake

unto him and showed him his great folly in this behalf, who yielding unto his fault prayed Master Doughty to be good unto him and keep it from the General [Drake]; but to be brief, he told him he could not keep it. . . .

So at the General's next coming aboard the prize, Master Doughty opened the same unto him, who presently falling into some rages, not without some great oaths, seemed to wonder what Thomas Doughty should mean to touch his brother, and did . . . assure himself that he [Doughty] had some farther meaning in this, and that he meant to shoot at his [Francis Drake's] credit, and he [Francis Drake] would not . . . suffer it.

From this time forth grudges did seem to grow between them from day to day. . . . Then was Master Doughty put again into the *Pelican*.

Doughty himself was accused by John Brewer, Edward Bright, and others of having taken goods from the captured *Mary* for his own use.

Having thus in the beginning of February put off the Island of Cape de Verde, we had not the sight again of any land until the 6th of April that we fell with the coast of Brazil. But in this mean season you shall understand what befell . . . Master Doughty. . . .

[One day, in the Atlantic] it chanced John Brewer, the trumpet[er], to go aboard the *Pelican*, where . . . the company offered him a [pie], among the which Master Doughty putting in his hand, said "Fellow John, you shall have in my hand, although it be but light amongst the rest"—and so laying his hand on his [Brewer's] buttock.

Which perceived of John trumpet[er], he began to swear wounds and blood to the company to let him loose, for they are not all (quoth he) the General's friends that be here. And with that turned him to Master Doughty and said unto him . . . "God's wounds, Doughty, what dost thou mean to use this familiarity with me, considering thou art not the General's friend?"

[Doughty] answered him: "What, fellow John, what moves you to this and to use these words to me, that am as good and sure a friend to my good General as any in this fleet, and I defy him that shall say the contrary? But is the matter thus, why yet, fellow John, I pray thee let me live until I come into England."

Thus John Brewer, coming again presently aboard the prize, had not talked any long time with the General, but the boat went aboard and rested not, but presently brought Master Doughty to the prize's side, Master Drake being in the midst of service. Who [Drake] hearing the boat at the ship side, stood up, and Master Doughty offering to take hold of the ship to have entered, quoth the General:

"Stay there, Thomas Doughty, for I must send you to another place"—and with that commanded the mariners to row him [Doughty] aboard the flyboat, saying unto him it was a place more fit for him than that from whence he came. But Master Doughty, although he craved to speak with the General, could not be permitted.

The flyboat *Swan* served as a storeship. Putting Doughty there was a disgrace, like locking him in the pantry. Drake soon sent other dissatisfied gentlemen there to keep him company.

A MYSTERIOUS FOG

The trans-Atlantic voyage was not all quarrels and conflict. In the "burning zone" of the equator, the seamen were frightened by "flashing lightnings and terrifying . . . claps of thunder." Flying fish and the "ravening fowl" which preyed on the fish fascinated the crews.

However, when a deadly fog blanketed them off São Pedro do Rio Grande do Sul, Brazil, and almost caused the ships to run aground there (April 5, 1578), many, including Drake, were ready to blame the well-educated Doughty for somehow causing the mist. Northern sailors had long attributed storms to sorcerers; Drake began saying that the bad weather "came out of Tom Doughty's cap case [hatbox]."

The following selection, from Francis Fletcher's manuscript narrative (Vaux, W. S. W., ed., *The World Encompassed*), describes the fog and gives the pilot da Silva's explanation of it.

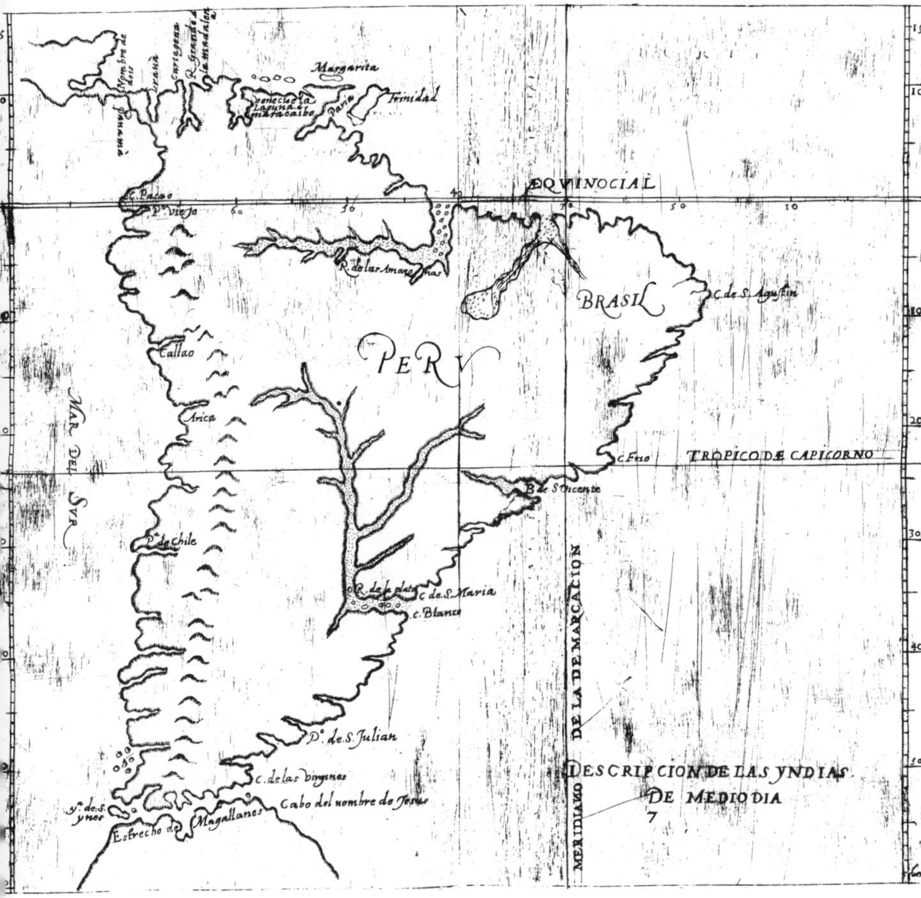

A map of South America made in 1601. It shows San Julian on the east coast; the Strait of Magellan; and Callao, the port of Lima, on the west coast.

We had not [long] held our way inward but the sight of land was taken from us . . . with such a haziness as if it had been a most deadly fog, with the palpable darkness of Egypt, that never a ship could see another. . . . [Then] did follow such extreme storms as [if] heaven and earth had gone together . . . and that which was a sign of a desperate state . . . we were upon a lee shore [shore toward which the wind was blowing], and the shoals increased upon us.

So that if the Portuguese pilot had not been appointed of God to do us good, we had perished . . . for he being well acquainted with the bloody government of the Portuguese, was not ignorant of this part of country [Brazil] and . . . he presently cried [to] return as we could. . . . One of our ships touched with the shoals, but by God's Providence came clear away, and being cast about to the seas . . . our fleet was . . . separated. . . .

Now the pilot being in the Admiral [flagship] . . . [explained] that such was the tyranny of the Portuguese towards the natural inhabitants [the Indians], that . . . they willingly exile themselves and . . . dwell in the . . . remote and unfruitful parts of the land. . . . [But the Portuguese] must pursue . . . them [there]. . . .

Wherefore this people . . . were now driven to yield themselves into the hands of devils, and took them for their . . . protectors against . . . the Portuguese. . . . When they [the Indians] see any ships upon their coasts, the shore being sandy, they cast the sand up into the air, whereof ariseth suddenly such a haziness as a most gross and thick fog . . . that the land cannot be seen, no nor the heavens.

. . . Withal [come] such horrible . . . winds, rains, and storms, that there is no certainty of life one moment of time. . . .

By this means did they continually overthrow the Portuguese . . . and . . . supposing us to be Portuguese . . . they did practice against us as against them.

SOUTHWARD BOUND

After the fleet escaped the squall, there was a question to be decided about da Silva. "All this month [March] he [Drake] ran along the coast of Brazil whither he went for to go to Espírito Santo to put me on land and give me my ship," wrote da Silva in his logbook. "And for this they held a council."

If there *was* a peace party, its members probably urged that the kidnapped pilot be freed, but instead both the expert da Silva and his caravel were kept.

Now winter approached in the Southern Hemisphere. The men suffered from the cold, and tempers grew short. By May 18, the fleet had reached a harbor in southern Argentina which Drake called Port Desire. During the two weeks they spent there they encountered friendly but thievish Indians, one of whom playfully snatched Drake's "scarlet sea-cap" off his head.

Just as they were about to set sail from an island in the harbor, a boatswain named Cuttill waded off toward the mainland crying "I will remain here with the cannibal Indians rather than accuse Thomas Doughty falsely!"

Although Cuttill later changed his mind, fired a shot, and was brought back to the fleet, he had supplied further evidence of the violent antagonisms in the expedition.

Indians in Argentina snatch Drake's cap from his head.

THE TRIAL OF THOMAS DOUGHTY

After leaving Port Desire, Drake navigated nearly to the Strait of Magellan, but then icy gales drove the fleet back north. On June 20, 1578, he entered the harbor of San Julian, about 150 miles above the Strait, where he decided to establish winter quarters.

It was a gray bitter-cold day. The *Pelican*, *Elizabeth*, *Marigold*, and da Silva's caravel anchored off a sandy island at the entrance to the bay. The gentleman-adventurers huddled together and talked in low voices the sailors could not overhear. Everyone stared from the ships through the gloom at the mainland and saw on the shore a decaying gallows.

It was the gallows on which, fifty-eight years earlier, Ferdinand Magellan had executed a mutineer when half his men had revolted against him there. Drake's cooper later made beer tankards out of the wood of this gallows, but not many wanted to drink from them.

Huts were built on the island, the seamen doing the work while the gentlemen watched, and Drake had Doughty closely guarded. Several of the ship captains cast sympathetic glances toward Doughty. Doughty's words to one of these captains, John Chester, had been reported to Drake. While a captive on Chester's *Swan*, Doughty had said to him: "Master Chester, let us not be thus used at these knaves' [*i.e.*, sailors'] hands. Lose nothing of that authority that the general [Drake] hath committed unto you. If you will, we will put the sword again into your hands and you shall have the government."

Two days later, June 22, 1578, when Patagonian Indians shot and killed two of Drake's men on the mainland, the atmosphere became even more tense. Seamen cursed gentlemen, and gentlemen scorned seamen.

Drake could see his entire expedition foundering on

this bleak coast. He imagined Doughty and his followers accusing him of being a madman who would lead them to their deaths; he could see Doughty enlisting the support of the captains who shared his class feeling, risking all to "put the sword" into their hands. Then they would sail off on their own venture, not the Queen's.

Ten years earlier, when he had attacked Rio de la Hacha on John Hawkins's "troublesome voyage," Drake had demonstrated that he was one who fired first and asked questions afterward. Now, suddenly, he brought Thomas Doughty to trial.

The trial was probably illegal. Drake claimed to have a commission from the Queen giving him power of life and death over members of the expedition, but no one ever saw this document. During the trial, Doughty admitted revealing the plan of the voyage to Burleigh against the Queen's express command, and Drake seized on that as treason. But that was only a technicality.

The real issue was who was to "have the government"—Drake, or the discontented men who opposed him. On June 30, 1578, Drake ordered everyone ashore on a rocky island in the harbor. He impaneled a jury of twelve, with one captain who had supported Doughty as foreman and another as prosecutor. But once the trial was decided on, Doughty was doomed. A verdict of Not Guilty would have been a verdict against Drake—who now named himself judge.

The only account of this famous trial is by Cooke, the seaman who later returned to England with Captain Winter. Some scholars consider Cooke prejudiced against Drake, but all authorities accept the speeches and actions he reports here as authentic. The following selection is from John Cooke's "Narrative" (Vaux, W. S. W., ed., *The World Encompassed*).

The last day of June, the General [Drake] himself being set in place of judgment (and having the whole company brought a shore) . . . the General spake unto Master Thomas Doughty . . . brought thither more like a prisoner than a gentleman of honest conversation . . . :

"Thomas Doughty, you have here sought . . . to discredit me, to the great hindrance and overthrow of this voyage, besides other great matters where with I have to charge you . . . the which if you can clear yourself of, you and I shall be very good friends, whereto the contrary you have deserved death."

Master Doughty answered, it should never be approved that he merited any villany towards him.

"By whom," quoth he [Drake], "will you be tried?"

"Why, good General," quoth he [Doughty], "let me live to come into my country, and I will there be tried by her Majesty's laws."

"Nay, Thomas Doughty," quoth he [Drake], "I will here impanel a Jury . . . to inquire farther of these matters. . . ."

"Why, General," quoth he [Doughty], "I hope you will see your commission be good."

"I warrant you," answered he [Drake], "my commission is good enough."

"I pray you let us then see it," quoth he [Doughty]. . . .

"Well," quoth he [Drake], "you shall not see it; but well, my masters, this fellow is full of prating, bind me his arms, for I will be safe of my life. . . ."

So they took him and bound his arms behind him; then he [Drake] gave divers furious words unto Thomas Doughty. . . .

Then . . . was there a Jury called, whereof Master John Winter was foreman. Then by John Thomas were these articles read unto them . . . all which Doughty did not greatly deny, until at length came in one Edward Bright . . . who said: "Nay, Doughty, we have other matter for you yet that will a little nearer touch you. . . ." . . . Then John Thomas read farther . . . that Thomas Doughty should say to Edward Bright in Master Drake's garden, that the queen's majesty and council would be corrupted. . . .

"Why, Ned Bright," said Master Doughty, "what should move thee thus to belie me? . . ."

Then it fell out upon farther talk that Master Doughty said that my lord Treasurer [Burleigh] had a plot [plan or route] of the voyage.

"No, that hath he not," quoth Master Drake. The other [Doughty] replied that he had.

"How?" quoth Master Drake.

"He had it from me," quoth Master Doughty.

"Lo, my masters," quoth he [Drake], "what this fellow hath done, God will have his treachery all known, for her Majesty gave me special commandment that of all men my lord Treasurer should not know it, but . . . his [Doughty's] own mouth hath betrayed him."

So this was a special article against him [Doughty] to cut his throat. . . . Then Master Doughty offered him if he would permit him to live and to answer these objections in England, he would set his hand to what so was there written. . . .

"Well once let these men," quoth he [Drake], "find whether you be guilty in this or no, and then we will farther talk of the matter." . . .

Then Master Leonard Vicary, a . . . friend unto Master Thomas Doughty, said unto him [Drake], "Why . . . there is, I trust, no matter of death."

"No, no, Master Vicary," quoth he [Drake].

So with this the Jury went together, finding all to be true. . . . Thus having received . . . the verdict, he [Drake] rose . . . and departed towards the water side, where calling all the company with him, except Master Thomas Doughty and his brother, he there opened a certain bundle of letters and bills and looking on them said, "God's will I have left in my cabin that I should especially have had" (as if he had there forgotten his commission . . .)

But here he showed forth first letters that were written . . . to my lord of Essex for his entertainment, secondly he showed letters of thanks from my lord of Essex . . . [praising] him . . . then read he letters from my lord of Essex unto secretary Walsingham in his great commendation . . . and lastly he read a bill of her majesty's adventure of a thousand crowns. . . .

So when he had all done he said more, "My masters

. . . you may see whether this fellow hath sought my discredit or no, and what should hereby be meant but the very overthrow of the voyage, as first by taking away of my good name . . . and then my life. . . .

"And now my masters consider what a great voyage we are like to make, the like was never made out of England, for by the same the worst in this fleet shall become a gentleman, and if this voyage go not forward, which I can not see how possibly it should if this man live, what a reproach it will be, not only unto our country but especially unto us. . . . Therefore my masters they that think this man worthy to die let them with me hold up their hands, and they that think him not worthy to die hold down their hands."

At the which divers that envied his [Doughty's] former felicity held up their hands, some other again for fear of his [Drake's] severity sticked not to lift their hands although against their hearts. . . .

[Drake then came] to his former judgment seat [and] pronounced him [Doughty] the child of death. . . . And said farther if any man could between that and the next morrow devise any way that might save his [Doughty's] life he would hear it. . . .

"Well, General," quoth he [Doughty], "seeing . . . you would have me made away, I pray you carry me with you to . . . Peru and there set me a shore."

"No truly, Master Doughty, I can not answer it to her Majesty if I should so do. But how say you, Thomas Doughty, if any man will warrant me to be safe from your hands and will undertake to keep you . . . see what I will say . . ."

Master Doughty then . . . [asked Master Winter to do this, and Master Winter agreed].

Then Drake a little pausing said, "Lo, then, my masters . . . we must . . . nail him close under the hatches and return home again without making any voyage, and if you will do so say your minds."

Then a company of desperate bankrupts that could not live in their country without . . . spoil . . . cried, "God forbid, good General!" . . .

Thus willing Master Doughty to prepare for his death . . . he [Drake] rose and departed.

On July 2, 1578, Doughty and Drake received the sacrament together at a religious service. Then Drake gravely entertained his old friend at a banquet. And then Doughty was beheaded.

Drake "made the head to be taken up and showed to the whole company."

"Lo, this is the end of traitors," said Drake.

GENTLEMEN AND MARINERS

Drake's execution of Doughty was like a major operation. Immediately afterward, the expedition was in a worse state than before, and Drake himself could not yet know whether the removal of Doughty would effect a cure.

There was deep gloom. Doughty's younger brother, John, "walk[ed] . . . solitary and mournful" along the shore of their island base. One of Doughty's accusers, Bright, quarreled with John and struck him. The gentlemen still huddled together, fearful rather than defiant now.

Sickness spread among the crews; rations ran short, and they ate mussels and seaweed to keep alive. Snow, sleet, and howling gales descended upon them from the Antarctic.

When on the Sunday following Doughty's death Drake "commanded the whole company to be ready to receive [communion] . . . saying that he would have all old quarrels . . . to be forgiven," the men received the sacrament, but numbly. The depression continued, and then Drake decided to tackle the problem of morale head-on.

On August 11, 1578, he had all members of the expedition assemble once more on the shore. He had a vision—he had a dream, not only of Spanish gold and silver lining the pockets of his men, not only of treasure galleons waiting to be plundered, but of Englishmen sailing the great South Sea for their Queen. If mariners and gentlemen would only pull together, they could do it! Out of tragedy could come a new unity.

The following selection, from John Cooke's "Narrative" (Vaux, W. S. W., ed., *The World Encompassed*), contains the famous sermon Drake preached on this occasion—a sermon which helped turn the tide of discontent and set the expedition on the road to achievement.

But now the 11th of August, he again commanded the company to be a shore, for that he had some matter of importance to say unto them. . . . Upon his coming a shore [he] entered into a tent, one side of the which was laid open, the company the better to see and hear what might be said; and he calling Master Winter on one side of him and John Thomas on the other side, his man laid . . . before him a great paper book, and withal Master Fletcher offered . . . to make a sermon.

"Nay, soft, Master Fletcher," quoth he, "I must preach this day myself." . . .

Then commanded he every ship's company severally to stand together, which was also done. Then said he,

"My masters, I am a very bad orator . . . but what so I shall here speak, let any man take good notice of . . . for I will speak nothing but I will answer it in England, yea and before her majesty. . . .

"Thus it is, my masters, that we are very far from our country and friends, we are compassed in on every side with our enemies. . . . Wherefore we must have these mutinies and discords that are grown amongst us redrest. . . . Here is such controversy between the sailors and the gentlemen, and such stomaching [taking offense] between the gentlemen and sailors, that it doth even make me mad to hear it.

"But, my masters . . . I must have the gentleman to haul and draw with the mariner, and the mariner with the gentleman. What, let us show ourselves all to be of a company. . . . I would know him that would refuse to set his hand to a rope, but

I know there is not any such here; and as gentlemen are very necessary for government's sake in the voyage, so have I shipped them for that, and . . . though I know sailors to be the most envious people of the world, and . . . unruly . . . yet may not I be without them.

"Also if there be any here willing to return home let me understand of them, and here is the *Marigold*, a ship that I can very well spare. I will furnish her to such as will return . . . but let them take heed that they go homeward, for if I find them in my way I will surely sink them. . . . I must needs be plain with you. I have taken that in hand that I know not in the world how to go through with all, it passeth my capacity. . . ."

Well, yet the voice was that none would return, they would all take such part as he did.

"Well, then, my masters," quoth he, "came you all forth with your good wills or no?"

They answered, that they came all with their wills.

"At whose hands, my masters, look you to receive your wages?"

"At yours," answered the companies.

"Then," quoth he, "how say you, will you take wages or stand to my courtesy?"

"At your courtesy," quoth the companies, "for we know not . . . what wages to ask."

Then he commanded the steward of the *Elizabeth* presently there to lay down the key of the room, which he did. Then turning . . . unto Master Winter, he said, "Master Winter, I do here discharge you of your captainship of the *E.*; and you, John Thomas, of the *Marigold*; and you, Thomas Hood, of your mastership in the *Pelican*; and you, William Markham, of the *E.*; and Nicholas Antony, of the *Marigold*; and, to be brief, I do here discharge every officer of all offices whatsoever." . . .

Drake discharged all of his officers to show that he had the power to do so—that all were dependent upon him. He accused at least two gentlemen of being as traitorous as Doughty, but said he would show mercy. His purpose was to overawe the company.

"And now, my masters," quoth he [Drake], "let us consider what we have done, we have now set together by the ears [upset, or disturbed] three mighty princes . . . her Majesty [Queen Elizabeth, and] the Kings of Spain and Portugal. And if this voyage should not have good success, we should not only be a scorning or a reproachful scoffing stock unto our enemies, but also a great blot to our whole country forever. . . ."

And so restoring every man again to his former office he ended, thus showing the company that he would satisfy every man or else he would sell all that ever he had unto his shirt. "For," quoth he . . . "it is only her Majesty that you serve and this voyage is only her setting forth." So wishing all men to be friends he willed them to depart about their business.

On August 17, 1578, Drake departed from what Fletcher called the "*Bloody* Island and Port Julian." At stops along the coast, Drake had broken up his smaller vessels for firewood. Now the remaining three ships, the *Pelican*, the *Elizabeth*, and the *Marigold*, manned by crews many of whom were still dispirited, sailed south toward one of the most dangerous stretches of water in the world: the twisting, storm-haunted 360-mile passage of the Strait of Magellan.

Chapter Four

THROUGH THE STRAIT

[We] saw fires on both sides
of the Strait and there were difference of opinion . . .
whether the land . . . on the [south] side . . .
is an island or main land.

—JOHN DRAKE

Four

THE GOLDEN HIND

Below San Julian, the coast of South America curves in and then back out, like a playground slide, to a tip named Cape Virgins after St. Ursula and her eleven thousand martyred maidens. A ship sailing past the pampas of southern Argentina must make almost a U-turn around the cape to enter the wide green waters of the Strait of Magellan.

On August 20, 1578, Drake reached Cape Virgins. He waited several days for a favorable wind before entering the Strait. While waiting, he shrewdly changed the name of his flagship from the *Pelican* to the *Golden Hind*.

A hind is a female red deer—but, more important, a golden hind was Sir Christopher Hatton's crest in heraldry. Sir Christopher Hatton, one of Drake's chief backers, had become Elizabeth's new favorite not long before Drake left England. If there were to be unpleasant repercussions from the execution of Doughty, it might be just as well for Drake to have Christopher Hatton and his golden hind in his corner!

The following selection, from *The World Encompassed* (1628), describes Drake's arrival at the Strait and his changing the name of his flagship.

August 17, we departed out of this port, and . . . set our course for the Straits, Southwest.

August 20, we fell with the cape, near which lies the entrance into the strait, called by the Spaniards *Cabo Virgen Maria*, appearing 4 leagues before you come to it, with high and steep gray cliffs, full of black stars, against which the sea beating, sheweth as it were the spoutings of whales . . . like cape Vincent in Portugal.

At this cape, our general caused his fleet, in homage to our sovereign lady the Queen's majesty, to strike their top-sails . . . and withal in remembrance of his honorable friend and favorer, Sir *Christopher Hatton*, he changed the name of the ship which himself went in from the *Pelican* to be called the *Golden Hind*.

Which ceremonies being ended, together with a sermon . . . we continued our course on into the said [sea], where passing with land in sight on both sides, we shortly fell with so narrow a strait, as carrying with it much wind, often turnings, and many dangers, requireth an expert judgment in him that shall pass the same: it lieth West North West and East South East. But having left this strait a stern, we seemed to be come out of a river of two leagues broad, into a large and main sea.

THE BOTTOM OF THE WORLD

As Drake navigated through the bleak Strait, he had to rely upon a sketchy account of Magellan's voyage and upon his seaman's intuition. By 1578 there had been five Spanish expeditions there, but Spain had no wish to attract visitors to this doorway to the South Sea and the riches of Peru and had therefore not encouraged the publication of information about the Strait.

It rained frequently. There were squalls of driving hail. Winds roared down from the snowy mountains, shifting suddenly from one point of the compass to another. Currents foamed side by side through the tortuous Strait in opposite directions, like trains passing.

But Drake and his men were pleased by the sight of beech and pine forests on the mountin slopes, and by the innumerable penguins, two thousand of which they killed for food supplies. Their spirits began to improve.

The following selection, from Francis Fletcher's manuscript narrative (Vaux, W. S. W., ed., *The World Encompassed*), describes the first part of Drake's journey through the Strait.

In passing alongst we plainly discovered that the same *terra australis* [southern land] . . . to be no Continent (and therefore no Strait), but broken Islands and large passages amongst them. . . .

In these Islands we found great relief and plenty of good victuals, for infinite were the number of fowl . . . named Penguin. . . . This fowl cannot fly, having but stub wings, without feathers, covered over with a certain down . . . as are also all their body besides; in their heads, eyes, and feet, they be like a duck, but almost as a goose. . . .

It is not possible to find a bird of their bigness to

have greater strength than they; for our men putting cudgels into their earths to force them out, they would take hold of them with their bills and would not let go their hold . . . and [the men] . . . could not in long time draw them out. . . .

We departing from these Islands had . . . a hard passage. . . . First, the mountains being very high . . . did every one send out their several winds; sometimes behind us, to send us on our way; sometimes on the starboard side, to drive us to the larboard, and so the contrary; sometimes right against us, to drive us farther back in an hour than we could recover again in many. But of all others this was the worst, that sometime two or three of these winds would come together . . . whose forces . . . did so violently fall into the sea, whirling, or as the Spaniard saith, with a *tornado*, that they would pierce into the very bowels of the sea, and make it swell upwards on every side. . . .

Besides this, the sea is so deep in all this passage, that . . . there is no coming to anchor. Neither may I omit the grisly sight of the cold and frozen mountains rearing their heads . . . into the cold and frozen region, where the . . . sun never reacheth to dissolve the ice and snow. . . . From these hills distilled so sharp a breath, that it seemed to enter into the bowels of nature, to the great discomfort of the lives of our men.

The trees of the Islands near . . . these mountains feel the force of the freezing streams which descend from them; for the snow which falleth upon them, and the rain which cometh down, do both freeze as they light upon the trees, and with their . . . weight . . . the main . . . boughs are crushed down so close together, that no art . . . can make closer . . . arbours than they be; under the which . . . is engendered such temperate heat, that the herbs may seem always to be green.

INDIAN NOMADS

The Strait of Magellan is not a comparatively short passageway, like the Panama Canal. It is longer than the distance from New York to Richmond, Virginia, a 360-mile ocean river which took Drake over two weeks to navigate. He and his men sailed along observing the shores on either side, like Huck Finn voyaging down the Mississippi.

Toward the western end of the Strait, they encountered four of the primitive Fuegian Indians—tall savages with great black eyes and coarse black hair, but more timid than the Patagonian giants of San Julian.

These Indians lived a nomadic life along the shores, catching fish and killing penguins for their food and for the rank-smelling oil they rubbed on their bodies. They carried glowing embers of permanent campfires in the bottoms of their boats, which had rounded ends. They are called Fuegian from the name Magellan gave the south coast of the Strait, along which he had seen mysterious night fires, *Tierra del Fuego* ("Land of Fire").

The following selection, from Francis Fletcher's manuscript narrative (Vaux, W. S. W., ed., *The World En-*

A map of the Strait of Magellan, published in Europe in 1598, fancifully pictures the Indians of the area. Because the Spanish kept their maps of the New World secret, Drake had to rely on written accounts of Magellan's voyage and his own seaman's intuition when he sailed through the Strait.

compassed), describes these Indians and their scanty household furnishings.

The Islands . . . were . . . frequented by a comely and harmless people, but naked . . . whom we could not perceive to have either set places or dwelling, or any ordinary means of living, as tillage, breeding of cattle, or any other profession, but [were] wanderers from place to place. . . . They builded little cottages of poles and boughs, like arbours in our gardens in England, wherein they themselves for the time lodge and keep their household stuff.

The value whereof [*i.e.*, of the household goods] would not . . . [amount to] xxv shillings in our country, for proof whereof I took an inventory of all the particulars of . . . the chiefest Lord's house . . . as followeth:

1 water pail.	2 wooden spits, and
2 drinking cups.	1 pair of racks.
2 boxes of stuff to paint.	2 hatchets, 1 knife.

1 fair floor of earth for a bed to lay upon without any clothes. . . .

Their water pails, drinking cups, and boxes, are made of the barks of trees sewed together with threads of the guts of some beasts, like lute strings. Their hatchets and knives are made of mussel shells. . . .

They grind them by great labour to a fine edge and very sharp, and as it seemeth, very durable. They set their hatchets in helves made of writhen rods. . . . With these and their knives they cut their poles and boughs. . . .

Touching their boats, they being made of large bark instead of other timber, they are most artificial and are of most fine proportion, with a stern and foreship standing semicircular wise. . . . With these boats, they travel from place to place among the Islands, carrying every man his family. In all our travels in any nation, we found not the like boats at any time for form and fine proportion. . . .

The people, men and women, are gentle and familiar to strangers, and paint their bodies with forms and divers colours; the men making red circles about their eyes and red strokes upon their foreheads for the most part, and the women wear chains of white shells upon their arms, and some about their necks, whereof they seem to be very proud.

"THE ROARING FORTIES"

On September 6, 1578, the *Golden Hind*, the *Elizabeth*, and the *Marigold* entered the South Sea—the Pacific Ocean—which Drake had first glimpsed from the "great high Tree" in Panama and vowed to sail on.

Drake was exultant. He wanted to place a monument to Queen Elizabeth on Cape Deseado ("Cape Desired"), at the western end of the Strait—Cape Deseado, in 52°

south latitude, balances London, near 52° north latitude—but a gale prevented his landing. So he studied the big chart he had obtained from Lisbon and pointed the prow of the *Golden Hind* northwest.

Drake's map showed an "elbow" of Chile jutting out into the Pacific the way Spain juts out into the Atlantic from Europe. But of course there is no such "elbow"—Chile slants back north-northeast. Drake's northwestern course would have taken him to Siberia, not Lima. Later Chaplain Fletcher decided the map must have been deliberately drawn wrong by the Spaniards to fool intruders into the South Sea.

On September 7, before Drake had sailed very far northwest and while he was still happy over having passed the dangerous Strait of Magellan, the skies clouded. Suddenly one of the "black storms" of the "roaring forties"—the high latitudes of the southern hemisphere—sent his tiny fleet reeling toward the Antarctic. For weeks darkness covered the heavens, the howling tempest battered the ships, and death stood beside each seaman.

Drake fought to keep the stern of the *Golden Hind* turned toward the mountainous waves. Since his map showed Terra Australis as a westward extension of Tierra del Fuego, he feared that at any moment this lee shore might loom out of the darkness and the fleet crash like egg shells on its rocks.

Drake was driven scores of miles south. Then early in October the storm lessened, and the *Golden Hind* and *Elizabeth* struggled back to a bay just north of Cape Deseado. But within an hour of their arrival the storm burst again. The *Golden Hind* was blown back south, this time among the islands below the Strait of Magellan. On October 19, 1578, Drake at last found shelter behind what Fletcher calls the "utmost Island . . . to the Southward of America"–probably Henderson Island, in 55° south latitude.

The following selections, from *The World Encompassed* (1628) and Francis Fletcher's manuscript narrative therein (Vaux, W. S. W., ed.), describe the great storm which left Drake with only one ship to continue the voyage.

The sixth of *September* we had left astern of us all these troublesome Islands, and were entered into the South Sea, or *Mar del Sur,* at the cape whereof [Cape Deseado] our General had determined with his whole company to have gone ashore, and there after a sermon to have left a monument of her Majesty, engraven in metal, for a perpetual remembrance. . . . But neither was there any anchoring, neither did the wind suffer us by any means to make a stay.

[We now set] our course to steer along the coast toward the Equinoctial [the equator] again. [But] following the directions of the common maps of the Spaniards we were utterly deceived, for . . . they had set forth the map false, that they might deceive strangers . . . that they might perish by the running off to the sea, rather than touch with any part of . . . America; for where the land trendeth . . . to the North East . . . they, in their maps, have laid it out to the Northwest. . . .

[Then] the wind [blew] right against us and so

violent . . . we were enforced back again with the lee shore of the Southerly Islands from whence we [had] departed, with a fearful look for destruction that night before day.

The day being come, the light and sight of sun and land was taken from us, so that here followed . . . a palpable darkness . . . of 56 days, without the sight of sun, moon, or stars, the moon only excepted, which we see in eclipse the space of a quarter of an hour or there abouts.

About which time, the storm being so outrageous and furious, the bark *Marigold*, wherein Edward Bright, one of the accusers of Thomas Doughty, was captain, with 28 souls, were swallowed up with the horrible . . . waves, or rather mountains of the sea. Which chanced in the second watch of the night, wherein myself and John Brewer, our trumpeter, being in watch, did hear their fearful cries, when the hand of God came upon them.

We thus deprived of our hand maid, continued without hope . . . in the violent force of the winds . . . and the . . . beholding . . . of the cragged rocks and . . . monstrous mountains, being to us a lee shore, where unto we were continually driven by the winds. . . .

Every mountain sent down upon us their several intolerable winds, with that horror that they made the bottom of the seas to be dry land . . . sending us headlong upon the . . . swelling waves . . . over the rocks. . . . At the last, in this miserable state we were driven as through the eye of a needle into a great . . . bay, by a most narrow passage of rocks, somewhat to the Northward of the Southerly Cape [Cape Deseado] of America, where . . . coming to anchor . . . being night [October 7], we had like entertainment from the hills as we had before from the mountains. . . .

Our cables broke, our anchors came home, our [two remaining] ships were separated. . . . We in the Admiral [flagship, the *Golden Hind*] were fully persuaded the Vice Admiral [the *Elizabeth*] was perished; and the Vice Admiral had the same opinion of us. . . . Neither did we see one the other again any more till we met together in England, for the Vice Admiral [the *Elizabeth*] was enforced . . . into the same passage, the . . . strait of Magellan [and returned home]. . . .

Notwithstanding, when it seemed good to God, after so long trial of . . . 56 days . . . [we arrived] at the utmost Island . . . to the Southward of America, whereat we . . . made [found] both the seas to be one . . . and that there was no farther land beyond the heights [latitude] of that Island, being to the Southward of the Equinoctial 55° and certain minutes . . . but that the way lay open for shipping . . . being the main sea. . . .

But before our going to land, we had a strange and sudden accident, for John Brewer, our trumpeter, standing upon the poop sounding his trumpet, being now as great a calm as it had been a storm, without any wind to move or shake a silken thread, most strangely a rope was so . . . violently hurled against his body that it cast his body over into the sea . . . eight times his length distant from the [ship] . . . where labouring mightily for life (the boat being not ready) many ropes were cast round about him . . . but he could not catch hold of any . . . till he called one by name to cast one to him, which no sooner was done, but he received it, and was saved at the last pinch.

It is doubtful that Fletcher and others on the *Golden Hind* could have heard "fearful cries" from the *Marigold* during the storm or even been certain that the *Marigold* sank then. Afterward Drake hoped he would meet both the *Marigold* and the *Elizabeth* in the Pacific. One scholar conjectures that the *Marigold* simply deserted and was later lost.

As for the *Elizabeth*, Captain Winter took her into the Strait, seeking shelter from the tempest. Then either he or his crew—accounts differ—decided to return to England. Winter claimed that headwinds made it impossible for him to sail back out into the Pacific.

THE MISSING CONTINENT

Where was Terra Australis? Where was the southern continent identified with mysterious Ophir of the Old Testament, from which King Solomon's ships brought gems and gold?

Drake concluded that it did not exist. The storm had shown him that it did not extend west of the Strait because if it did, the *Golden Hind* would have been wrecked against it. And Drake suspected that Terra Australis would not be found in the south Atlantic either.

"If a man be furnished with wood and water, and the wind good, he may keep the Main Sea, and go round about the Straits to the Southwards," said Sir Richard Hawkins some years later, basing his statement in part on what "Sir Francis Drake told me." Thus Drake revised his preconceived ideas about geography in the light of experience–unlike Columbus, who believed until the day of his death that the islands and mainland he had discovered were part of Asia.

Drake also boasted to Hawkins that he had been farther south than any man in the world. For in the "Southermost Island"–probably Henderson Island–Drake had gone ashore, sought out "the Southermost part of the Island, cast himself down upon the uttermost point groveling, and so reached out his body over it." When Drake returned to the *Golden Hind*, he told his men that he had been farther south "than any of them, yea, or any man as yet known!"

Chaplain Fletcher erected a monument at this same point but, neglecting to hang himself over the cliff like a flagpole, did not get as far south as his commander.

Of course the Horn Islands (55°, 58′) are a few miles farther south than Henderson Island (55°, 36′). This group was discovered in 1616 by the Dutchman Schouten and his companion Le Maire; they named the cape farthest south Van Horn. English writers immediately argued that Drake's "Southermost Island" *was* Cape Horn, but most scholars now reject the claim. The topography and Indians described by Fletcher differ greatly from those of the Horn Islands.

The following selection, from Francis Fletcher's manuscript narrative (Vaux, W. S. W., ed., *The World Encompassed*), reflects Drake's common-sense reasoning about the union of the two oceans and the nonexistence of Terra Australis.

In this Island were growing wonderful plenty of . . . currants. . . . Myself being landed, did, with my bag, travel to the southern most point of the Island . . . where I found that Island to be more Southerly three parts of a degree than any of the rest of the Islands. Where, having set up on end a stone of some bigness, and with such tools as I had . . . engraven Her Majesty's name, her kingdom, the year of Christ, and the day of the month, I returned again . . . to our company.

We departing hence and taking our farewell from the southernmost part of the world known . . . we altered the name of those Southernly Islands from *Terra Incognita* [Unknown Land] . . . to *Terra Nunc Bene Cognita* [Land Now Well Known], that is, Broken Islands; which in coasting it again . . . returning to the Northward, we proved to be true. . . . But before I proceed any farther . . . I think it not impertinent here to . . . resolve that great error, which hath been and is maintained of many touching the Strait of Magellan—videlicet, that it is a . . . strait indeed. . . .

The chief . . . [reason for this belief] is that the

Southerly land is a continent, but two continents made a strait, therefore it is and must be a strait.

I answer, it was ever uncertain, from the first discovery of that passage by the Spaniards, and could not be determined by Magellan himself, that that land was a continent, but left it under the name of *Terra Incognita* [Unknown Land]. And what others afore or since have written or said . . . are but guesses and imaginations . . . until now we have . . . put it out of doubt to be no continent . . . but broken Islands . . . and therefore no strait.

Their second reason is . . . [that] there runneth such a current between the land of America and this *Terra Australis* [Southern Land], (caused, say they) by the continent, that it carieth ships headlong through to the South Sea, but admitteth no return back again; which could not be, if there was any other . . . passage for the sea to keep its natural course. . . .

That this is false, is evidently proved by experience, as well . . . from Magellan himself, as from ourselves; for at what time he passed that way, being come into the South Sea, one of his ships stole away and returned into their country the same way they came. The like thing chanced to ourselves, for . . . our Vice Admiral [the *Elizabeth*] was enforced into the supposed strait again with the violence of the storm, and so passed the whole way we had come without any such hindrance, yea, they had more sweet and quiet return back again, than we had a-coming forward. . . . To make an end of this matter, the reason why the Spaniards have abused the world with such an untruth . . . is, that thereby they would prevent all other nations . . . to travel that way, being (as they feigned) so impossible a thing to have any return, but must be forced to compass the whole world to their country again.

And thus much for the clearing of . . . doubt . . . and . . . to make it known to the world that there is neither continent, current, nor strait.

Fletcher's definition of a strait as used in the previous selection was that "two continents made a strait." Since it had then been discovered that there was no second continent, he argued that there was no strait.

AMBUSH ON MOCHA

The weather cleared at last. There was bright sunshine, and a cold wind blew from the south. Drake gave the command to raise sails.

On October 28, 1578, as the *Golden Hind* moved slowly away from the "Southermost Island," which Drake had named Elizabeth Island, Drake counted his blessings. His flagship, at least, had survived the storm. He hoped they would find the *Elizabeth*, perhaps also the *Marigold*, at the rendezvous at 30° south latitude. And here he was, sailing the great South Sea in a heavily armed ship, completely unsuspected by the Spaniards!

Drake started again on a northwest course, but after a few days, not sighting land, he shifted to northeast. As soon as he picked up the coast he tried another northwest "zig," but once more had to "zag" back northeast. Thus for much of the first part of the voyage he was far out at sea, missing the islands and archipelagos that fringe southern Chile.

Near the end of November, returning to the Chilean coast for the second time, Drake sighted the Valdivia River. He then decided to navigate just offshore and, on November 25, arrived at Mocha Island, eighteen miles west of the coast. Here Drake planned to obtain food and water before seeking the treasure ships of Peru.

But for the first and almost the last time on his daring venture, Drake was off guard. Southern and central Chile were wilderness—he saw no signs of Spaniards yet. What he did not know was that he had nevertheless reached a bloody frontier—the *Tierra de Guerra* ("Land of War"), where Spaniards and fierce Araucanian Indians fought each other without quarter.

Spaniards were safe only in fortified towns, like settlers in our West. The Indians hated them—and Spaniards were the only white men the Indians knew.

The following selection, from Francis Fletcher's manuscript narrative (Vaux, W. S. W., ed., *The World Encompassed*), tells what happened when Drake sought supplies from these Indians.

In our course from Elizabeth Island we kept close aboard the shore . . . of the . . . mainland . . . till we fell in sight of an Island named Mocha. . . .

Arriving at Mocha, the General with a chosen company rowed to land . . . [where] the inhabitants repaired to them with great show of friendship, and . . . entered into traffic with our men . . . to our great good liking. For their commodities were such as we wanted, as fat muttons, hens, maize. . . . And because we had need of fresh water . . . they gave us assurance by signs . . . the next morning . . . we should have it at pleasure. . . .

The captain [Drake], the time being come, made such choice [of men] as he thought fit . . . who together with joy set forwards to land. . . . Being . . . well armed with sword and targets [shields], they feared no perils, nor made no doubt but to have as kind entertainment now as the night before—but . . . they had neither bow nor other shot. . . .

A narrow creek was appointed for them [Drake's men] to bring their boat into, on both sides whereof did grow abundance of Indian reeds, high and thick, wherein they [the Indians] had couched a multitude of bloody soldiers on both sides, not one whereof made any show, at their [the sailors'] coming in, but lay close. . . .

Some were appointed to stand at the landing place to receive our men, and . . . to show them a place where to go for water, for which purpose two of our men carrying vessels with them went on shore, one whereof took the boat's rope with him to draw the boat nearer to land, which done he left it laying on the ground. Which was no sooner out of his hand but an Indian took it up, holding the boat fast to the shore by it. Which done, others laid violent hands

Mocha Island, eighteen miles west of the coast of Chile. Drake put in to this island to get supplies but was attacked by Indians.

upon our men which were landed and carried them away.

The soldiers hidden in the reeds, well armed with bows, arrows, and darts, made of canes . . . set upon . . . them which were in the boats, who being as naked, not either able to defend themselves or annoy their enemies, were enforced to be butts to every arrow . . . so that not any person escaped without some grievous wound, and most had many, so that their bodies were laden with arrows from 2 to 3, 4, 5, 8, 10, and one had 21. . . . Some [were shot] in head and face, as the General [Drake]; some in the throat, breasts, arms, back, belly, and where not.

Neither had any one of them escaped with life, if one of the simplest of the company had not with his sword cut the boat's rope, and by that means set them at liberty. . . . At whose departure, arrows were sent to them so thick as gnats in the sun, and the sides of the boat within and without stuck full of them. . . . Who coming on ship board, the horror of their bloody state wounded the hearts of all men to behold them. Notwithstanding . . . it pleased God not any but one died of that accident.

Now in the mean time . . . the boat was manned again, and sent to see if happily any help might be had for our men whom the enemy had taken, but all in vain. . . . For when our men came in view of them, the multitude was great, by estimation 2000 persons . . . with bows, darts, spears, shields, pikes, and other weapons, most of them headed with pure silver, which in the light of the sun made a wonderful . . . glittering.

Among these . . . were our two [captured] men in their . . . torments. . . . The men being fast bound, were laid upon the ground . . . and the people cast themselves into a ring round about them, hand in hand with a dance, still turning . . . with a song. In the meantime tormentors, working with knives upon their bodies, cut the flesh away by gobbets, and cast it up into the air, the which falling down, the people caught in their dancing, and like dogs devoured it in most monstrous and unnatural manner [until the captives died]. . . .

Our men . . . after a volley or two of shot bestowed upon them, but could do them no harm because at every offer of a shot they would all fall flat to the ground . . . returned again as men all-amort [half dead] with so horrible a spectacle. . . .

We might have taken a revenge upon them . . . with great shot out of our ship, but the General would not . . . consent to it. We understood that this people . . . did inhabit at mainland; but by the bloody cruelty of the Spaniards (as were the Brazilians by the Portuguese . . .) were driven to . . . this Island . . . where . . . they maintain [themselves] with continual shedding of the blood and eating of the flesh of the Spaniards. . . .

And because they were never acquainted with any other nation but the Spaniards . . . persuading themselves that we were the same, and the rather because one of our men rashly spake in the Spanish tongue, they determined to bestow upon us a Spaniard's reward.

The Araucanians were the Apaches of South America—a fierce, independent people who maintained their freedom until the last half of the nineteenth century. They are the heroes of a famous Spanish epic, Ercilla's *La Araucana*, written just before Drake's arrival. With his two head wounds, Drake was fortunate to escape with his life from Mocha.

Chapter Five

RAIDING AN EMPIRE

It was notorious, the harm that had been perpetrated in the ports . . . from . . . Chile to this . . . City of . . . [Lima] by a ship of English Corsairs. . . . It is necessary to . . . chastise and annihilate them.

—VICEROY AND ROYAL *Audiencia* OF PERU

Five

THE ENGLISH PIRATE

In 1577 a comet had blazed through the Pacific skies and been feared as an omen of future evil by the peoples of Chile and Peru. Only a year later Drake arrived; he now lurked just off Chile, like a sea dragon reconnoitering the peaceful South American coast.

Chile was a poor, undeveloped country, without the gold Drake sought. Settlers, led by the conquistador Valdivia, had only come there from Peru in 1540. Towns were still clusters of huts, and there were few harbors along the sandy shore.

On December 5, 1578, Drake entered a port of nine households, which would later become the great city of Valparaiso. This was the first notice the Spaniards had of the presence of "the English Pirate" among them; they were as astounded as if the *Golden Hind* had come from outer space.

The following selection, from *The World Encompassed* (1628), tells what happened in Valparaiso from the English point of view.

The same day that we received this dangerous affront [in Mocha], in the afternoon, we set sail from thence. . . . Falling with a bay called Philips Bay, in 32 deg. . . . *November* 30, we came to anchor, and . . . sent our boat to discover what . . . the place would offer. . . . Our boat . . . could find . . . [n]either . . . fresh victuals or . . . fresh water. . . .

Yet in their return . . . they descried within the bay an Indian with his canoe, as he was a fishing: him they brought aboard our general, canoe and all. . . . A comely personage, and of a goodly stature; his apparel was a white garment, reaching scarcely to his knees; his arms and legs were naked; his hair upon his head very long; without a beard, as all the Indians for the most part are. He seemed very gentle. . . .

[From him] we understood that there was no . . . way to have our necessities relieved in this place; so he offered himself to be our pilot to a . . . good harbor, not far back to the Southward . . . where, by way of traffic [trading], we might have . . . both water and . . . other things. . . . *December* 5, by the willing conduct of our new Indian Pilot, we came to anchor in the desired harbor.

This harbor the Spaniards call Valparaiso, and the town adjoining Santiago. . . . We found in the town diverse storehouses of the wines of Chile; and in the harbour, a ship called the . . . *Grand Captain of the South* . . . laden . . . with the same . . . liquors . . . [and] a certain quantity of fine gold . . . and a great cross of gold beset with emeralds, on which was nailed a god of the same metal.

We spent some time in refreshing our selves and easing this ship of so heavy a burthen, and on the 8 day of the same month, having . . . stored ourselves with . . . wine, bread, bacon, etc., for a long season, we set sail . . . carrying again our Indian pilot . . . whom our General [Drake] bountifully rewarded . . . and caused him . . . to be landed in the place where he desired.

How Drake's raid appeared to the Spaniards is narrated by Pedro Sarmiento de Gamboa, one of the greatest of Spanish navigators, who at this time was living in Lima. Sarmiento had not only been captain and chief pilot on Mendaña de Neyra's voyage of discovery to the Solomon Islands, 1567–69, but was also a historian of the Incas and a humanist whose free thinking got him into trouble with the Inquisition. From inhabitants of Valparaiso he collected information which he wrote up in a "Narrative," from which the following selection is taken (Nuttall, Zelia, ed., *New Light on Drake*).

From . . . [Mocha, Drake and his men] followed the coast of Chile and cast anchor at the port of Quintero, six leagues from the port of Santiago, and there they took an Indian from whom they learnt that they had left Valparaiso, the port of Santiago, six leagues behind them. Taking this Indian as a guide they entered the harbour of Valparaiso on Friday, December fifth, 1578, at noon and cast anchor in the middle of the bay. . . .

The Corsair sent the skiff with eighteen Englishmen, arquebusiers, archers and men with shields, to seize a merchant vessel that lay at anchor in the port and was named "La Capitana" [the flagship] because she had served as such in the voyage of discovery to the Solomon Islands. She was about to sail for Peru and had on board five sailors and two negroes.

The Englishmen entered her and took the sailors below deck and locked them up. Then some of them went to fetch their chief, the Corsair Drake, who went to "La Capitana" and placed guards in her.

Some men went ashore, and broke open the warehouses, thinking that they would find gold, but instead they only found wine, flour, salt pork, lard and suet. They took one thousand and seven hundred jars of wine and whatever else they wanted and transferred all to "La Capitana," in which they found twenty-four thousand pesos of gold. . . .

On Saturday, the sixth of December, at noon, the English Corsair set sail, taking "La Capitana" and the plunder with him. He placed twenty-five men in her to guard and navigate her, and on the following Sunday, in the afternoon, they cast anchor in the port of Quintero, six leagues from Valparaiso, and sent ashore the Indian they had seized.

The Englishmen took the sea-chart of the pilot of

"La Capitana" and guided themselves by it from port to port.

This "sea-chart" was valuable to Drake, not because navigation along Chile and Peru is difficult at this time of year—winds and currents are from the south, and there are no reefs—but because it depicted Spanish ports, where he might find a treasure ship.

From Quintero they came to the bay of Tanquey to take in water and not finding any proceeded to the port of La Herradura where both ships cast anchor and took in a supply of water and pigs. While on shore they heard an arquebuse shot inland and inferred that Spaniards were there.

The Englishmen on board stationed a sentinel in the crow's nest and at noon Drake came out of his cabin and perceived on land a horseman followed by about fifty or sixty others and by Indians. Francis signalled to his men on shore to withdraw to a rock that was near and to which they could wade.

Whilst they were doing so the Spaniards reached the shore and the last Englishman [Richard Minivy] who had remained behind to assemble the others was reached and killed by the Spaniards. But the Englishman was the first to discharge his arquebuse and he also drew his sword to defend himself.

When the Englishmen had reached the rock Francis sent the launch in which they all entered and went to the ship which immediately set sail.

The Spaniards beheaded Richard Minivy's body, then solemnly burned a book they found in his breast pocket, containing quotations from the Protestant Bible jotted in its margins. According to the Inquisition in Chile, the greatest danger from these pious pirates was not "their coming to disturb the peace and commerce . . . but the influence they exerted in spreading the ideas which they brought with them."

The Inquisition, which investigated Catholic Christians to see if they were guilty of heresy (holding wrong beliefs), functioned as a kind of secret police; it had several headquarters in the Spanish New World.

A SLEEPING SPANIARD

From La Herradura, Drake continued northward, looking for fresh water and for his two missing ships. Fresh water is scarce in central and northern Chile. The country behind the coast is a desert; the few small streams usually sink into the sand before they reach the sea.

Drake stopped in a deserted cove for a month to overhaul the *Golden Hind* and assemble a pinnace. Then on February 4, 1579, he went ashore in his pinnace to the tiny settlement of Tarapacá. Here he found a Spaniard asleep beside a pile of silver bars worth three thousand pesos.

This drowsy caballero was symbolic of all Spanish America. Viceroys, *audiencias*, harbor officials—all were taking a siesta which "the English Pirate" was about to disturb rudely.

The following selection, from *The World Encompassed* (1628), describes this and other incidents of Drake's voyage along the coast of Chile in the bright sunshine of the southern summer.

This [La Herradura, where Richard Minivy was killed] being not the place we looked for, nor the entertainment such as we desired, we speedily got hence again, and *December* 20, the next day, fell with a more convenient harbour, in a bay somewhat to the Northward. . . .

In this place we spent some time in trimming of our ship, and building of our pinnace. . . . But still the grief for the absence of our friends [the *Elizabeth* and the *Marigold*] remained with us, for the finding of whom our General [Drake] . . . intended . . . with his pinnace and some chosen men . . . to return back to the Southwards again, to see if happily he might . . . find them in some harbour or creek, or hear of them. . . . But after one day's sailing, the wind being contrary . . . he was forced . . . to return again.

Within this bay . . . we had such abundance of fish . . . that our gentlemen sporting themselves day by day with 4 or 5 hooks or lines, in 2 or 3 hours, would take sometimes 400, sometimes more. . . .

January 19, we set sail from hence. . . . As we

sailed along, continually searching for fresh water, we came to a place called *Tarapacá,* and landing there we lighted on a Spaniard who lay asleep, and had lying by him 13 bars of silver, weighing in all about 4000 Spanish ducats.

We would not (could we have chosen) have awaked him of his nap. But seeing we, against our wills, did him that injury, we freed him of his charge [*i.e.*, the silver], which otherwise perhaps would have kept him waking, and so left him to take . . . the other part of his sleep in more security.

Our search for water still continuing, as we landed again not far from thence, we met a Spaniard with an Indian boy, driving 8 lambs or Peruvian sheep [llamas]: each sheep bore two leathren bags, and in each bag was 50 pound weight of refined silver, in the whole 800 weight.

We could not endure to see a gentleman Spaniard turned carrier so, and therefore without entreaty we offered our service and became drovers, only . . . we could [not] keep the way which he intended; for almost as soon as he was parted from us, we with our new kind of carriages [*i.e.*, the sheep and the silver] were come unto our boats.

LLAMAS AND SILVER

The *Golden Hind* now arrived at the coast of Peru—Peru with its desert wild flowers and "desert drums" (the sound made by grains of sand eddying in the heated atmosphere). Here was the lofty valley of the Cuzco, where the Inca "people of the sun" had created their splendid civilization.

Some of Drake's gentlemen, gazing at the wall of the Andes, which seemed to hold up the sky, might have read about the bards and dramas of the Incas, their gigantic stone temples unequaled in Europe, their roads and post-houses, rituals, and accurate statistics kept by the *quipus*, or system of knots.

Here also were llamas and the silver mines of Potosí. Drake's seamen stared wide-eyed at the Peruvian "sheep" —the llamas, members of the camel family, are accurately described in the selection to follow. As for Potosí, ever since the "Silver Rush" began there in 1545, this settlement had been a household name the world over.

In that year, on the bleak plain 13,000 feet above the sea, an Indian, or the Indian's llama, pulled up a bush and disclosed an incredible vein of silver. Now a city of 100,-000 stood in this sky-wilderness, a city of adventurers, gamblers, and millionaires, like California's later '49ers—for in Cuzco, theater tickets sold for fifty dollars apiece and Spanish beauties paid fourteen thousand dollars for a dress for a fiesta.

The snow-capped peak above Potosí was now honey-combed with mine shafts. In half a century, by slave labor of the Indians, 396 million dollars worth of silver was taken from Potosí. In a century and a half, the bullion of Potosí would triple the silver supply of Europe.

The port of Arica, where the silver was loaded on ships to be carried to Panama, drew the *Golden Hind* like a magnet. The following selection, from *The World Encompassed* (1628), tells what happened there from the English point of view.

In Peru, llamas were used to transport gold and silver.

A sixteenth-century European artist's naive version of the silver mines of Potosí in Peru. Indian slaves mined the silver for the Spanish.

Amongst other things . . . the sheep [llamas] of the country (viz., such as we mentioned before, bearing the leathren bags) were most memorable. Their height and length was equal to a pretty cow, and their strength fully answerable. . . . Upon one of their backs did sit at one time three well grown and tall men, and one boy, no man's foot touching the ground . . . the beast nothing at all complaining of his burthen. . . .

These sheep have necks like camels, their heads bearing a reasonable resemblance of another sheep. The Spaniards use them to great profit. Their wool is exceeding fine, their flesh good meat, their increase ordinary, and besides they supply the room of horses for burthen or travel. Yea they serve to carry over the mountains marvelous loads, for 300 leagues together, where no other carriage can be made but by them only.

Hereabout, as also all along, and up into the country throughout the Province of *Cuzco*, the common ground . . . in every hundred pound weight of earth, yieldeth 25 s. of pure silver. . . .

The next place likely to afford us any news of our ships . . . was the port of . . . *Arica*, standing in 20 deg., whither we arrived the 7 of *February*. This town seemed to us to stand in the most fruitful soil that we saw all alongst these coasts . . . in . . . a most pleasant and fertile valley. . . . It hath continual trade of shipping, as well from Lima as from all other parts of Peru. It is inhabited by the Spaniards.

In two barks here we found some forty and odd bars of silver (of the bigness and fashion of a brickbat, and in weight each of them about 20 pounds), of which we . . . ease[d] them, and so departed. . . .

In our way to *Lima,* we met with another bark at *Arequipa,* which had begun to load some silver and gold, but having had (as it seemed, from Arica by land) some notice of our coming, had unloaded the same again before our arrival.

Nuño da Silva said that although Drake "went ahead in the pinnace and used oars and sail, he found the vessel already emptied. He [Drake] learned that . . . the silver had been landed two hours previously, and the people on shore shouted at him, saying 'Go, you thief!' and jeered at him, saying 'that he had lost the prize by two hours!' " (Nuttall, Zelia, ed., *New Light on Drake*).

Yet in this passage we met another bark loaded with linen, some of which we thought might stand us in some stead, and therefore took it with us.

The raid on Arica is described from the Spanish point of view in the following selection, from Nicolas Jorje's "Deposition before the Royal Court of Panama" (Nuttall, Zelia, ed., *New Light on Drake*). Nicolas Jorje was a Flemish seaman on one of the ships captured by Drake.

I was in the ship of Felipe Corço, in the port of Arica, Peru, wherein there also lay at anchor a ship which had come from Lima laden with merchandise, and with about three hundred jars of wine. At vesper-time there arrived in the same port an English vessel whose Captain was an Englishman named Captain Francis Drake, who carried much artillery, many arquebuses and other arms. . . . The English ship was accompanied by another which . . . had been taken by the Englishman in the port of Santiago de Chile.

The said Englishman also brought a launch filled with armed men and forthwith seized the ship on which I was. Taking possession of her by placing some of his men in her, he seized thirty-five bars of silver, alloyed with white mercury, and a chest which belonged to men who had come from Potosí and contained small pieces of silver. . . . He burnt another ship which was with him after taking out the wine she contained and transferring it to the ship on which I was. . . .

He took me with him by force . . . and threatened to kill me many times. . . . For he said that I had deceived him by not informing him that in the port of Arequipa there was a vessel laden with much silver.

When the said Englishman was on his way to Arequipa, he sent his launch to search for the port, and she returned with a ship which she had found in . . . Arequipa. I saw how the English crew of the launch . . . told their captain that they had found [the ship] empty but that it was evident that . . . her cargo had been removed. They thought that she must have contained much silver, for there were many persons on shore. . . .

Then the said Englishman [Drake] realised that the inhabitants of the port of Arequipa had received a warning from those of Arica, and hurried on to reach Lima. In order to do this he abandoned, about eight leagues off the coast, in sight of land, the three ships he had taken. . . . Before leaving them . . . he took from them much wine and other things, but left much wine and much timber in them. He hoisted the sails . . . and let them drift without any crew. I do not know what happened to them.

A NIGHT ATTACK

Disappointed at Arequipa—the sides of the Spanish silver ships had still been wet above the water line when he arrived—Drake now continued northward.

Thus far he had combined caution with his daring. He had retreated at La Herradura, losing only the reckless Minivy. After seizing the ships at Arica, Drake had bombarded the town during the night in preparation for a landing—but in the morning, seeing mounted Spaniards on shore, he abandoned the attempt.

On February 13, 1579, Drake captured the ship of one Gaspar Martin at sea. Martin told him there were two ships laden with silver at Callao, and that another, under San Juan de Anton, had just left Callao for Panama. Then Drake decided to throw caution to the winds.

"The worst in this fleet shall become a gentleman!" Drake had promised his men at Doughty's trial. Here was their golden opportunity.

Callao was the port of Lima, chief city of Peru and seat of the viceroy. In Lima, Drake learned from his Spanish captives, John Oxenham was a closely guarded prisoner of the Inquisition. Oxenham's audacious attempt to seize the Isthmus of Panama with fifty men had ended in disaster in 1577—according to legend because of his love for a Spanish beauty. Could Drake now rescue his old friend as well as enrich his men?

Drake sailed swiftly toward the capital. Lima's white towers rose in a setting of green, churches, monasteries, schools, palaces—and barracks of soldiers. At Callao there would be not only a number of Spanish ships at anchor, but also harbor launches and customs officials to investigate strangers from the sea.

But Drake did not hesitate. That night, standing beside a Portuguese pilot he had taken from Gaspar Martin's ship, Drake peered into the darkness and strained his ears for sounds other than the slap of waves against the *Golden Hind*'s prow—as the pirate ship and its pinnace slipped into Callao harbor.

The few stars of the Southern Hemisphere shed little light. A high rocky island loomed ahead—then the ship's keel jarred against a shoal.

The following selection tells what happened next. It is by a young cousin of Drake's who sailed with him and who bore the same name as the brother Drake had lost in Panama—John Drake (from "Second Declaration before the Inquisition in Peru," printed in Nuttall, Zelia, ed., *New Light on Drake*).

In view of the fact that she [the ship of Gaspar Martin] did not carry any silver or gold . . . [the English] left her, taking only a Portuguese mariner whom they understood to be the pilot of the bark. Captain Francis [Drake] took him into his galleon with the object . . . that the pilot should take them into the port of Callao.

On sailing between the main land and the island we struck a shoal and thought that we had run aground. Believing that the Portuguese had done this maliciously, Captain Francis threatened to cut off his head.

However we proceeded towards the island and entered the port [at 10 P.M.] simultaneously with another vessel. We anchored close together and when the Englishmen inquired whence she had come, the other ship answered "from Panama."

Captain Francis ordered the Spaniards he carried on board to say that his was a galleon from Chile; then in the boat, with six or seven men, accompanied by the pinnace, carrying twenty or thirty men, he went to the other vessels anchored there and cut their cables. . . . This was done so that . . . the wind would carry these ships out of the port, where he could seize them, and hold them for ransom, so that, in exchange, they would give him the English Captain who was said to be a prisoner in Lima. . . . This Captain's name was John Oxenham.

But the wind fell and the ships remained motionless and, not having found either gold or silver therein we went in the pinnace to the ship from Panama and attacked its crew, but were not able to take her. One Englishman named Thomas was killed.

Because her crew had offered resistance Captain Francis went to his galleon and fired a cannon-shot at the ship from Panama, which went through both of her sides without killing anyone. Seeing this, those of the ship abandoned her and went ashore in their boat.

Captain Francis sent in pursuit of the boat, in order to take her but was not able to do so. One of her men, however, who seemed to be a half breed, threw himself into the sea and he was taken into the galleon. The Englishmen then seized the Panama ship which had been abandoned.

Suddenly a sound of great shouting came from the shore and the bells began to ring. A custom house inspection boat came out, and when she came alongside of the galleon, and the occupants perceived her cannon they began to shout "Frenchmen! Frenchmen!"

All that night we lay becalmed but the tide carried us out of the port. On the next day we saw two or three vessels and a launch come out of Callao in pursuit of us.

On seeing this Captain Francis, imagining he might have to fight, summoned to his galleon the men he had placed in the Panama ship. He left in her the sailor he had carried from Chile, named Juan Griego [John the Greek]; as well as the Fleming he had taken in Arica and the half-breed from Callao.

Then he went his way.

Drake had cut the cables of two large vessels so they were adrift and had taken one captive ship, "the Panama ship," with him. The Panama ship contained no treasure, and he abandoned it just outside Callao, probably to slow up his pursuers, who turned aside to retake it.

Did Drake try to rescue Oxenham? The selection above says Drake planned to hold the vessels whose cables he cut as ransom in exchange for Oxenham. But not all modern historians are convinced of this; Drake's actions seem to have been more directed toward acquiring gold than toward saving his friend. And he really did not have enough men to attempt a rescue.

Drake did later send several messages to Viceroy Don Luis de Toledo, threatening reprisals if Oxenham and the three other English captives were harmed—but in vain. In October, 1580, Oxenham was hanged in Lima.

THE CHASE—I

Don Luis de Toledo, Viceroy of Peru, was an energetic veteran who had established a system of forced labor among the Indians and had put to death Tupac Amaru, the last of the Inca chieftains. When he was roused from a sound sleep at 1 A.M. and told that "the English Corsair" was in Callao, he was furious.

"To arms!" shouted Don Luis as torches flared through the dark streets.

The viceroy quickly covered the six miles between his palace and Callao. At his command, sailors and gentlemen embarked immediately in two ships, in pursuit of the pirate. Later, a fleet would be dispatched to stand guard at the Strait of Magellan to intercept Drake on his return.

But Don Luis, in his anger and confusion, overlooked the queen on the ocean chessboard—the galleon *Nuestra Señora de la Concepción*, nicknamed the *Cacafuego*. Captain San Juan de Anton, in the *Cacafuego*, was sailing unsuspectingly north with a cargo of gold and silver valued at several million dollars. And Don Luis did not think to send a courier by land to San Juan de Anton's first port of call, Paita, to warn him of Drake's presence.

The following selection by one of the pursuers narrates the Spanish chase after Drake; it is from the "Narrative of Pedro Sarmiento de Gamboa" (Nuttall, Zelia, ed., *New Light on Drake*).

On Friday, February thirteenth, 1579, between ten o'clock and midnight, the ship of some English Corsairs, with a pinnace and skiff arrived at the port of Callao de Lima. Entering between the ships that lay at anchor there, the Corsairs enquired for the ship of Miguel Angel, for they had learnt that many bars of silver had been embarked on her. On boarding her they found however that . . . the silver had not yet been carried aboard.

They then went in the pinnace and skiff from vessel to vessel. They cut the cables of seven of the nine vessels that were lying at anchor there, so that they should drift and not be able to follow them.

When they reached the ship of Alonso Rodriguez Baptista, which had just arrived from Panama with a cargo of Castilian stuffs, they boarded her, shooting many arrows at her sailors and pilot. The said Alonso Rodriguez was wounded by an arrow and it is said that one Englishman was killed.

The Englishmen seized the ship with all her cargo, and carrying her with their ship, pinnace and skiff, set sail around the island of the port towards the northwest. They were able to do this in safety because the inhabitants of the coast by which they had come had not made haste in sending information to the Viceroy.

While the English Corsairs were plundering the ship, the sailors who had escaped went on land giving the alarm. As soon as this was received . . . the royal officers and the chief governor mustered the people so as to repel the Corsairs [Drake's men], and sent dispatches to the Viceroy who was in Lima . . . and who received them one hour after midnight. . . .

His Excellency [the Viceroy Don Luis de Toledo], with great dispatch, armed himself and ordered the gentlemen and his retainers to do the same. The bells were rung to give the alarm and criers were sent from door to door to explain what had occurred and to summon all inhabitants to assemble in the public square, where His Excellency took his stand. . . .

When the entire population had assembled in the square His Excellency mustered them and, as many had no arquebuses, he had the armoury opened and distributed many arquebuses, pikes and provisions of ammunition.

While this was being done he received information that the Corsairs . . . were English. Until then it had not been known for certain to what nation they belonged. . . .

Thereupon he sent the General Diego de Frias Trejo with his men to defend the port of Callao and guard the King's money which was about to be embarked and consisted of more than two hundred thousand pesos in bars of silver. The General departed and reached Callao in all possible haste, but the English Corsair was already far out at sea, although he could still be seen from land, taking with him the ship of Alonso Rodriguez Baptista.

As the General considered it advisable to go after the Corsair . . . two vessels were designated, one being that of Miguel Angel. As soon as the men arrived from Lima they embarked in both ships. Three hundred men . . . thus embarked, all with a great desire to chastise the Corsair, just as though it were each man's private affair.

The General embarked in Miguel Angel's ship,

which thus became the *Capitana* (or flagship). Pedro de Arana embarked as Admiral . . . in the other (which became the *Almiranta* . . .), and thus we sailed after the Corsairs who were already more than four leagues out of the port, towards the north-west.

When the *Capitana* came under the lee of the island, the wind fell and she was detained a long while, whereas the *Almiranta*, which followed, overtook and passed ahead because as she sailed further from the island her sails caught more wind.

It was afterwards known that when the English Corsair saw vessels coming out of the port he inquired of the Spaniards on his ship what vessels could they be? They answered that it seemed to them that they were some ships coming after the vessels whose moorings he had cut. . . . When the Englishmen saw this they took great care, nevertheless, to watch what course we were taking.

When they saw that we were adopting the same as theirs and that we had already cleared the island and reached the open sea, the Corsair Francis realised the truth. Dissembling, he said to the sailors whom he had taken prisoners that he would now release them according to his previous promise.

He ordered them to enter the merchant ship which he had seized and to return in her to the port. He sent them to the merchant ship in the pinnace and gave order that the Englishmen who had been working that vessel were to return in the said pinnace.

As these Englishmen delayed in returning . . . he jumped into his skiff and went to the vessel, wrangling with his men. These all jumped into the pinnace and went to their ship. . . . The merchant ship, with four or five of the liberated sailors, returned towards the port.

After collecting his men, the Englishman spread his top-gallant sails and took flight towards the northwest. . . .

Our *Almiranta* . . . followed the Englishman for a whole day. At sunset she was almost out of sight, and had gained headway because our vessels were without ballast, and, being crank, could not bear sail and pitched with the moving about of the men, thus navigating very slowly. Moreover the English ship, being further out at sea, caught a stiffer breeze and sailed on the wind.

Notwithstanding all this, although at dusk we lost sight of her, we did not give up pursuing her during a great part of the night. And those of the *Almiranta*, having sent to the merchant ship for a mariner named Juan Griego [John the Greek], who had come with the Englishman from Chile, ascertained from him that the English vessel was large and strong and carried seventy-five or eighty men and many pieces of artillery, beside many fire-instruments.

Having learnt this Pedro de Arana and several of those on the *Almiranta* went during the night to the *Capitana* and held a consultation with the General and other gentlemen as to whether they should continue to follow the enemy or go back to the port so as to get reinforcements. . . .

It was the General's view that they should continue . . . but many held a contrary opinion . . . [because of] the defectiveness of the vessels and the fact that they carried no food whatever and not sufficient artillery and ammunition or fire-instruments to cope with the English. . . .

Our . . . ships, carrying no ballast, could not possibly catch up with the enemy. Even if they did so our men would certainly be injured by the enemy's artillery, for our ships carried nothing but arquebuses to use against him.

Moreover, the most imperative reason for returning seemed to be that many of the gentlemen were very seasick and were not in a condition to stand, much less to fight. . . .

Finally, at the end of much discussion, it was resolved to return in order to obtain reinforcements so as to sally forth again, better equipped to attack the enemy. Thus they returned.

The viceroy was so enraged that he would not permit the seasick caballeros to land, and he threatened to arrest the lot of them.

THE CAPTURE OF THE *CACAFUEGO*

Blissfully unaware of the disturbance in Callao, San Juan de Anton, captain of the *Nuestra Señora de la Concepción*, sailed past northern Peru (now Ecuador). He thought his galleon was carrying the treasure of Potosí to Panama, whence it would be transferred to Nombre de Dios and then taken to Seville.

The silver would become bullets. It would buy arms for the Duke of Alva's soldiers to use against Protestant rebels in the Netherlands.

But one Sunday—March 1, 1579—San Juan de Anton discovered that his silver was not going to Seville after all. It was going to London.

The following selection, from the "Narrative of Pedro Sarmiento de Gamboa" (Nuttall, Zelia, ed., *New Light on Drake*), is based on San Juan de Anton's own account of the capture of the *Cacafuego.*

At noon on Sunday, the first of March, San Juan de Anton, being out at sea in his ship, between the Cape of San Francisco and the point "de la Galera," saw, close to land, a ship which was going the same way, bound for Panama. He thought she was a bark from Guayaquil and bore towards her.

Drake apparently wished to capture the *Cacafuego* at night because he reduced his speed by dragging wine jars filled with water behind the *Golden Hind.*

At about nine o'clock at night, the English ship crossed the course of San Juan's vessel and, immediately, came alongside. San Juan saluted but the Corsair did not return the salute. Believing her to be a ship from Chile . . . Master de Anton came to the side. By that time the English were already grappling his ship shouting: "Englishmen! strike sail!" Some one said: "Strike sail, Mr. Juan de Anton; if not, look out, for you will be sent to the bottom."

San Juan answered: "What old tub is this [which gives me orders] for striking sail? Come on board to strike [the] sails yourselves!" On hearing this they blew a whistle on the English ship and the trumpet responded. Then a volley of what seemed to be about sixty arquebuses was shot, followed by many arrows, which struck the side of the ship, and chain-balls shot from a heavy piece of ordnance carried away the mizzen and sent it into the sea with its sail and lateen yard.

After this the English shot another great gun, shouting again "Strike sail!" and, simultaneously, a pinnace laid aboard to port and about forty archers climbed up the channels of the shrouds and entered San Juan de Anton's ship, while, at the opposite side, the English ship laid aboard.

It is thus that they forced San Juan's ship to surrender.

They inquired for the pilot and captain from the selfsame San Juan de Anton, who was alone on deck. He would not answer them. Not seeing any other person on deck, they seized him and carried him to the English ship where he saw the Corsair Francis

Drake, who was removing his helmet and coat of mail. Francis Drake embraced San Juan de Anton, saying: "Have patience, for such is the usage of war," and immediately ordered him to be locked up in the cabin in the poop, with twelve men to guard him.

On the following Monday morning, at nine o'clock, the Corsair [Drake] went to breakfast on San Juan's ship. He had, meanwhile, left orders with his chief sergeant to prepare his table for San Juan de Anton, as though it were for himself.

Francis Drake remained until noon on the captured vessel, examining the riches she carried, and returned to his ship in the afternoon. Departing from the point where he had taken the vessel, he sailed with a fair wind . . . towards the north-west on the route to Nicaragua. . . .

During the first three days of fair weather, he transferred, by means of the pinnace, all the silver from San Juan de Anton's vessel to his ship, keeping meanwhile, as prisoners on his Admiral's ship the Spaniards whom he had found on the plundered vessel, who were ten or eleven persons, five being sailors and the remainder passengers.

The registered silver seized amounted to 362,000 pesos in bars, reals and gold. Of this 106,000 belonged to His Majesty and the rest to private individuals. . . . With what was on board beside this, the total amounts to more than 400,000 pesos.

He took all the victuals he wanted and two casks of water, tackle, sails, canvas and a cable.

Francis Drake complained of the Viceroy of Mexico, saying that he had broken his word to John Hawkins and had not observed the King of Spain's warrant of safety. Francis Drake stated that he had been present [at San Juan de Ulua] and had lost seven thousand pesos in that defeat and that three hundred Englishmen had been killed.

He added that . . . [because] the King had, since that time, been his treasurer for the sum that had been taken from him ten years ago, he now wished to act as treasurer of the King's estate. Therefore the silver which he took from the King was for himself; the silver taken from private individuals was for his Queen. . . .

He said he cared nothing for the Viceroy of Peru nor for all his people. He charged San Juan de Anton to beg the Viceroy of Peru from him not to kill the English prisoners [John Oxenham and others], and said that if they were killed, it would cost more than two thousand heads . . . those of people . . . of these parts. . . .

San Juan de Anton told him that since the Englishmen had not been killed up to that time the Viceroy would hardly kill them then. The Englishman asked San Juan what he thought was the Viceroy's intention concerning them. He answered that they would probably be sent to Chile to serve as soldiers . . . against the Indians.

Francis rejoiced greatly on hearing this and became pacified; for he displayed much anger whenever he spoke about them. . . .

Before releasing San Juan's vessel, the Englishman made several gifts to those whom he had robbed. He gave thirty or forty pesos in cash to each. To some he gave pieces of stuff from Portugal, and agricultural implements, such as hoes and pruning-knives. . . .

To San Juan de Anton he presented a firelock saying that it had been sent him from Germany. . . . To the clerk he gave a steel shield and a sword saying that he did this so that the clerk might appear to be a man-at-arms.

To San Juan he [also] gave two casks of tar, six hundred-weights of iron from Germany, and a barrel of powder. To a merchant . . . he gave some fans with mirrors, saying that they were for his lady. And to San Juan de Anton he gave a silver-gilt bowl, in the centre of which his name, "Francisqus Draques," was inscribed.

The Englishman [also] showed San Juan de Anton a navigation chart of more than two rods in length, saying that it had been made for him in Lisbon and had cost him 800 cruzados. He said that there were four possible routes for him to take from this South Sea for England.

One was by the Cape of Good Hope and India; another by Norway [the Northeast Passage, above Russia]; another by the Strait of Magellan. He would not name the fourth. . . . He said that he thought of returning to England very shortly, in less than six months.

San Juan de Anton responded that he would not be able to return even in a year's time, because he was in a "cul de sac." The Englishman told him that . . . he was satisfied with his proposed route and was going to follow it.

San Juan de Anton says that he believes, beyond a doubt, that the Englishman is going by the coast of Nicaragua and New Spain [Mexico]. . . . For the Englishman told him that he was going to take in water at the island of Caño near Costa Rica because he had no more. . . .

On the following Saturday, March 7th, he [Drake] sent all the prisoners back to the plundered vessel and told San Juan de Anton that he could go where he willed. San Juan de Anton sailed to the north-east . . . but kept sight of Francis Drake for about two days and observed him sailing for a long time towards the north-north-west.

A LETTER FROM FRANCIS DRAKE

When Drake let the *Cacafuego* go, a Spanish ship's boy (*grumete*) shouted out, "Our ship shall no more be called the *Cacafuego*, but the *Cacaplata*. Your ship shall be called the *Cacafuego*"—which made Drake's seamen roar with laughter. According to some scholars, *Cacafuego* means "Spitfire"; *Cacaplata* would then mean "Spit-silver."

Drake was delighted with his luck and himself. He had

The Golden Hind *capturing the Spanish ship* Nuestra Señora de la Concepción. *Drake took several million dollars in gold and silver from the Spanish treasure ship.*

made the *Golden Hind* a floating treasure house, with deposits valued at several million dollars in its hold. He could well afford to play the great lord with his prisoners.

His irrepressible sense of humor got the better of him, and he called for the "Register" of the *Cacafuego*—the list of goods shipped, with the names of the owners opposite. One by one he checked off the shipments of gold and silver. Then he signed his name at the bottom. Francis Drake was the new owner!

Still, he did not forget the ships he had lost off Tierra del Fuego: the *Elizabeth* and the *Marigold*. Hoping they were in the Pacific, he gave San Juan de Anton a letter addressed to John Winter, captain of the *Elizabeth*. The letter contained information about Drake; it would also serve as a safe-conduct to keep San Juan de Anton from being robbed twice.

"I am doing you a great benefit," Drake solemnly told San Juan de Anton. "Captain Winter is a very cruel man. If he came across you he would not leave one of your men alive! But with this letter, you are safe."

The following selection is Drake's letter to Winter (Nuttall, Zelia, ed., *New Light on Drake*)—the only surviving writing of Drake's connected with this voyage.

Mr. Winter.

If it please God that by a favourable chance Your Honour should meet San Juan de Anton, I pray you to treat him well, in accordance to the word I have given him. If Your Honour should be lacking in any of the things that San Juan de Anton carries, pay him double their value in the merchandise that Your Honour carries. Give orders that none of your soldiers are to do him harm or wound him.

What we determined about the return to our country will be carried out if God so wills, although I greatly doubt whether this letter will reach your hands, I abide as God knows, constantly praying to the Lord who holds you and me and all the world in His keeping to save or to damn. I give him thanks always. Amen.

This my writing is not only for Winter but also Mr Thomas, Mr Charles and Mr Caube and Mr Anthony and all the other good friends whom I commend to Him who redeemed us with His Blood. I have faith in God that he will not inflict more toils upon us but will help us in our tribulations.

I beseech you for the love of Jesus Christ, that if God permits you to suffer afflictions you do not despair of the great Mercy of God, for the great Prophet says that the Lord grants and gives new life. May God thus have mercy and show his compassion—to Him be glory, honour, power and empire, for ever and ever, amen, amen.

I, the mournful captain whose heart is heavy for you,

FRANCIS DRAKE.

THE CHASE—II

Sailing steadily on a north-northwest course, Drake drew away from the *Cacafuego* and disappeared over the horizon. But he left all Chile and Peru in an uproar. The pirate must be caught and punished! But how? The *Golden Hind* was a needle in a blue haystack that spread halfway around the globe.

On February 27, 1579—a few days before Drake overtook the *Cacafuego*—two more ships outfitted by the angry viceroy had left Callao. They had strict orders to bring back "the English Corsair," Francis Drake.

As usual on Spanish ships, the command was divided between a general in charge of the hundred and twenty soldiers, and an admiral who governed the sailors. The sergeant major was the veteran Pedro Sarmiento de Gamboa.

Early in March, as Drake fled north after robbing the *Cacafuego*, Sarmiento gave excellent advice about where to look for him. Sarmiento guessed the rendezvous that Drake was sailing toward. The admiral enthusiastically accepted Sarmiento's theory, and Drake's days seemed to be numbered.

The following selection, from the "Narrative of Pedro Sarmiento de Gamboa" (Nuttall, Zelia, ed., *New Light on Drake*), describes the second pursuit of Drake.

As soon as His Excellency, with great sorrow, heard of the return of the ships [from the first pursuit of Drake] he immediately came from the city to the port. . . . He had two vessels fitted out, and . . . ordered one hundred and twenty soldiers to embark, beside the sailors. On Friday, February twenty-seventh, 1579, we embarked. . . .

All the men left with a great desire to fight the enemy but with little hope of being able to overtake him, for he already had a start of fifteen days. After going out to sea they began to parley with all ships they met and to run along the coast, exploring ports and points with a sailing launch that we took with us for this purpose.

When we arrived at Santa we learnt that the Corsair had passed by there a fortnight previously and that, beyond the port of Trujillo, he had seized the bark of a certain Cataro and had taken from her what he wished.

We therefore immediately started for the port of Trujillo, having also heard that six days previously a great vessel with a sprit-sail, which was believed to be the pirate-ship, was thereabouts. . . . In that vicinity we saw a sail, and . . . bore down upon her . . . but she turned out to be a merchant's bark.

Passing onwards we arrived at Payta [northern Peru] on March tenth and learnt that the Corsair had entered and left the port a fortnight previously. . . . He had not even cast anchor, for he had learnt that a pilot named San Juan de Anton had started ahead of him on the previous day from the same port, with many bars of silver. . . .

Following the coast we proceeded and arrived off the point of Santa Elena [southern Ecuador] on

March 13th. The General sent men in the launch to . . . obtain information, but there was not a single ship in the harbour, and a letter written by the innkeeper was found in which he stated that he was fleeing on account of the news about the Englishmen.

The prudent innkeeper may have heard the rumor that not one but three English pirate ships were now roaming the South Sea. "It is considered certain that . . . two English ships . . . said to have come . . . [with] Captain Francis [Drake] . . . had arrived [off Peru]," reported one of the captives Drake released. Mothers were making their children behave by warning that "Francisco Draque" would "get" them if they were bad.

A letter was left behind for him recording that our fleet had been there. From thence the fleet went to the port of Manta where we cast anchor on March 17th, and found two ships lying at anchor. One of these was that of Bravo who, on his way from Guayaquil to Panama, carrying gold, had been robbed by the Englishmen. . . . They had taken from him fifteen thousand pesos in gold . . . as well as all the clothing and food they wished from certain trunks.

The Corsair made experiments in sailing with the bark, but as she sailed faster on the wind than his own ship he wrapped her sails around her anchor and cast them into the sea so that she could not sail ahead and give warning.

He released the men and the bark, permitting them to go back, and giving them a little coarse linen to make the small sail, with which Bravo came to Manta.

These men related that like a shameless robber who fears not God or man, the Corsair made many arrogant speeches, saying that San Juan de Anton could not escape him.

Therefore the Admiral [Diego de Frias Trejo] and the Sergeant-Major [Sarmiento] agreed that it would be advisable, in order to make some headway in the pursuit of the Corsair, to cross over from Manta to Nicaragua. The Admiral proposed this to the General. . . .

Pedro Sarmiento stated that, for reasons of navigation, the Corsair could not escape any other way than by the coast of Nicaragua and New Spain [Mexico] and that he would not go to the Gulf of San Miguel [Panama] because he was acquainted with the fact that Pedro de Ortega was there with soldiers. Besides the Corsair knew what had happened there to other Englishmen [John Oxenham and his men] who had attempted to enter and leave by that route. Moreover he would not wish to lose his ship. . . . Neither this nor the silver could be carried on men's shoulders [across the Isthmus of Panama]. . . .

Sarmiento [also] argued that . . . the Corsair would [not] dare to return by the northern coast of

Peru [en route to the Strait of Magellan] because he had left all the country in tumult and up in arms. Besides . . . [he] must suspect . . . that so powerful a country must be backed by a fleet. . . .

[But] the Corsair will have learnt from the Portuguese pilot he carries that on all that coast [of Nicaragua] there are no Spanish settlements or Indians who could offer resistance. . . . Nor are there any ships which could follow him. Therefore he could . . . enter ports wheresoever he willed . . . and . . . could freely rob the few ships that trade with cacao . . . as well as some vessels coming from the Philippines laden with gold and articles of great value. For such things are greatly coveted by pirates. . . .

In two paragraphs omitted here, Sarmiento speculates that Drake will try to return to England by the still undiscovered Northwest Passage, which "is not unknown to the cosmographers"—that is, Drake will continue on up the coasts of Mexico and California until he reaches this passage.

[Thus] it would be an impossibility to overtake him [the Corsair] by following him along the coast to Panama. . . . The best way would be to cross over [to Nicaragua]. . . . The higher we would go the better, for if he had not arrived at the coast when we did we could return along the coast looking for him. . . .

All these conclusions met with the approval of all the men on the *Almiranta* and of many on the *Capitana*, more especially those who were seamen, all of whom volunteered with good will to undertake the said voyage.

Chapter Six

THE COAST OF CALIFORNIA

The great probability is . . . that the Corsair has wintered . . . in the region of the Californias, in those small bays or deep coves . . . there.

—GUATEMALAN OFFICIAL TO KING PHILIP II

Six

ESCAPE TO NICARAGUA

As the *Golden Hind* fled northward and the viceroy's galleons plowed steadily on in pursuit, Drake found his ship losing speed because her hull was coated with barnacles. His water casks were almost empty.

Therefore, Drake turned aside to the island of Caño, off the rocky coast of Nicaragua. He anchored in a bay behind the island on March 16, 1579, and hoped that his pursuers would sail past his hideaway.

On March 20, out at sea Drake saw a small Spanish bark voyaging south. Had the bark seen the *Golden Hind*?

"Man the pinnace!" ordered Drake, without hesitation. He could not risk a report to officials in Panama of his whereabouts.

The following selection, from the viewpoint of Colchero, a Spanish pilot who was a passenger on the bark, relates what happened next; it is from the "Deposition of the Pilot . . . Colchero" (Nuttall, Zelia, ed., *New Light on Drake*).

. . . [On March 20, 1579] I was on my way to the city of Panama by order of . . . the Viceroy of

New Spain [Mexico], in a bark belonging to Rodrigo Tello, which had sailed from the port of Costa Rica with ten or eleven men, passengers and sailors. Arriving on the said Friday at noon, close to the Island del Caño, a launch full of Englishmen came towards us.

This launch contained about twenty Englishmen, archers and arquebusiers, who took us prisoners, and led us before an Englishman whom they named "Francisco Draque," who was in a vessel with high sides, of about three hundred tons, which was anchored in a cove.

He had us put into his vessel and kept us there for five days, at the end of which he released the others, and sent them ashore in a launch. The said Englishman would not release me, saying that he had been informed that I was a pilot for the China route, and that he wished me to conduct him thither.

I answered that I was not familiar with that route and begged him not to take me for I was a poor man and had a wife and children. I . . . said that I was not a pilot, but only a sailor.

The said English captain thereupon answered that I was not to plague him by speaking such nonsense and that I had to go with him . . . and that he would hang me if I talked any more.

Colchero asked and received permission to write letters to the viceroy of New Spain (Mexico) and to his wife.

Suddenly, on the same day, the said Englishman set sail, and navigated along the whole of this coast of Nicaragua. When he arrived . . . close to this town [of Realejo], he tempted me, with many promises of silver and gold, to go with him to England and to become a Lutheran. . . .

When he saw that I showed no disposition to accept his offers he said: "You! You must be a devoted subject of your King Don Felipe [Philip II], and a great captain!"

Then he requested me to pilot his vessel into this port [of Realejo], through the bar, so that he could burn a vessel therein . . . which [he knew] . . . was being fitted out for the navigation of the China route. . . .

I answered that I had never entered this port and would therefore not dare to . . . pilot him into it. Then the said captain uttered many threats and promises. . . .

When the captain saw that I would not do it he gave orders that I was to be hanged. Twice they placed a rope around my neck and raised me from the ground. When they saw that I was exhausted, they left me alone.

When the said Englishman saw that he could not carry out his evil design, he continued his voyage. . . . When close to the volcanoes of Guatemala they met a frigate in which came Don Francisco de Zarate. . . . They kept this frigate three days. . . .

At the end of three days, the said Englishman

promised . . . to release me and he did so, and put me into the ship in which Don Francisco de Zarate also came. . . . I saw . . . that the Englishman directed his course towards the port of Guatulco. . . .

The said Englishman took from me all the letters I carried and my navigation charts. . . . They said that they were going to return straight to their country by the Molucca route.

From Plymouth to Lima, Drake had sailed with capricious Lady Luck at his side. Now she smiled on him again. Colchero and a brother pilot had been en route to Panama to conduct a new governor across the Pacific to the Philippines. They carried top-secret navigation charts for the great South Sea, of which Englishmen were totally ignorant. These charts fell into Drake's hands, and he had a return ticket to London.

Having learned the winds, the currents, and the sailing route from Acapulco to the Philippines, he could go that way. The Moluccas were only one stop beyond the Philippines, and then Drake's Portuguese map would direct him around the Cape of Good Hope.

Drake calked the *Golden Hind* down to the water line, then sailed north. On March 27, he decided that both his prisoners and his pinnace—the "destroyer" with which he ran down merchantmen—were an encumbrance. After removing his cannon from the pinnace, he ordered all captives except the obstinate Colchero into it and set them free to row ashore. But he kept the captured Spanish bark as a storeship.

Scanning the southern horizon day and night for sails of

the avenging Spaniards of Lima, Drake continued northward.

A POLITE PIRATE

Meanwhile the Spaniards on the pursuing fleet faced a critical decision. They were off Ecuador, where the coast curves east toward Panama. Should they take this turnoff, or should they go on northwest, straight across the Gulf to Nicaragua?

Pedro de Sarmiento and the Admiral, Diego de Frias Trejo, urged the General, Don Luis, son of the viceroy, to sail on to Nicaragua. His instructions would cover him, they pointed out; he was ordered to "pursue [the Englishmen] by sea and by land."

"I will take a resolution in this matter," said young Don Luis de Toledo, "tomorrow."

Tomorrow came, and the General argued against going to Nicaragua. He must protect the Isthmus against "the Corsair." After Drake's exploits of 1573, Spaniards were very nervous about treasure trains crossing to Nombre de Dios. However, there would be a final decision—tomorrow.

Tomorrow came once more, and the verdict was postponed until the next day. The next day, the General said "Tomorrow." Finally, on the fourth and final mañana, the General, young Don Luis, announced his resolution: The fleet would go to Panama.

It was March 21, 1579, the day after Colchero was captured. Drake had now escaped the viceroy of Peru—although Drake did not know it. (Neither did the viceroy—he was going to be furious.) However, Drake was about to knock the *Golden Hind* against a hornets' nest in

Mexico. He was going to enter ports for water, and he was going to attack Acapulco itself—so he said.

On April 4, 1579, Drake captured a Spanish vessel bound south for Peru. It carried linens, silks, and porcelain from China—and one distinguished passenger, Don Francisco de Zarate. Don Francisco was a cousin of the Duke of Medina-Sidonia, one of the most illustrious noblemen of Spain.

The following selection is Don Francisco de Zarate's account of his capture. It is from a letter Zarate wrote to Drake's old enemy, Don Martin Enriquez, who had trapped Drake and Hawkins at San Juan de Ulua (Nuttall, Zelia, ed., *New Light on Drake*).

I sailed out of the port of Acapulco on the twenty-third of March and navigated until Saturday, the fourth of April, on which date, half an hour before dawn, we saw, by moonlight, a ship very close to ours.

Our steersman shouted that she was to get out of the way and not come alongside of us. To this they made no answer, pretending to be asleep.

The steersman then shouted louder, asking them where their ship hailed from. They answered, "from Peru," and that she was "of *Miguel Angel*," which is the name of a well-known captain of that route. The spokesman on the ship was a Spaniard. . . .

The ship of the adversary carried her bark [the pinnace] at her prow as though she were being towed. Suddenly, in a moment, she crossed our poop, ordering us "to strike sail" and shooting seven or eight arquebuse shots at us.

We thought this as much of a joke as it afterwards turned out to be serious.

On our part there was no resistance, nor had we more than six of our men awake on the whole boat, so they entered with as little risk to themselves as though they were our friends. They did no personal harm to any one, beyond seizing the swords and keys of the passengers. . . .

They ordered me to go in their boat to where their general was—a fact that I was glad of, as it appeared to me that it gave me more time in which to recommend myself to God. But in a very short time we arrived where he was, on a very good galleon, as well mounted with artillery as any I have seen in my life.

I found him promenading on deck and, on approaching him, I kissed his hands. He received me with a show of kindness, and took me to his cabin where he bade me be seated and said: "I am a friend of those who tell me the truth, but with those who do not I get out of humour. Therefore, you must tell me (for this is the best road to my favour), how much silver and gold does your ship carry?"

I said to him, "None."

He repeated his question.

I answered, "None, only some small plates that I use and some cups—that is all that is in her."

He kept silent for a while, then renewing the conversation asked me if I knew Your Excellency [Don Martin Enriquez].

I said, "Yes."

"Is any relative of his or thing pertaining to him on this ship?"

"No, sir."

"Well, it would give me a greater joy to come across him than all the gold and silver of the Indies. You would see how the words of gentlemen should be kept."

I made no reply to this. He then stood up, and bidding me go with him, led me to a cabin situated in the poop below deck, where there was a prison. . . . In it, at its end, was an old man.

[Drake] said to me "I . . . want you to tell me who that man is in there."

I answered that I did not know him.

"Well," he said, "know that it is a pilot named Colchero, whom the Viceroy was sending to Panama to convey Don Gonçalo to China [the Philippines]." He then had the pilot released . . . and we all went up on deck. . . .

On the following day, which was Sunday, in the morning, he dressed and decked himself very finely and had his galleon decorated with all its flags and banners. . . . He said to me: "Let one of your pages come with me to show me your apparel."

He [then] went from his galleon at about nine in the morning [to our ship] and remained until towards dusk, examining everything contained in the bales and chests.

Of that which belonged to me he took but little. Indeed he was quite courteous about it. Certain trifles of mine having taken his fancy, he had them brought to his ship and gave me, in exchange for them, a falchion and a small brazier of silver, and . . . he lost nothing by the bargain.

On his return to his vessel he asked me to pardon him for taking the trifles, but that they were for his wife. He said that I could depart the next morning, when the breeze would rise, for which I gave him thanks.

The next morning, which was Monday, he gave back to some of the passengers . . . their boxes. . . . He ordered his sloop to be prepared and manned with two dozen archers. He had one of the artillery men called and ordered him to carry aboard half a dozen pieces of artillery.

This done, he told me to embark with him. . . . I did so, and on arriving at our vessel he boarded her first and having all our sailors called together, he gave each one a handful of reals. He also gave the same to some other men who appeared to him to be the most needy.

He commanded that one of those sailors should embark with him so as to show him where water was to be obtained. All excused themselves, saying that they did not know where water was to be had, so he caused Juan Pascual [one of Zarate's sailors] to be put by force in his sloop saying that he would hang him if he replied a word.

With this he took leave of me, and his last words were to beseech me, earnestly, to tell certain Englishmen who were in Lima that I had met him on April 6th and that he was well.

Drake now released Colchero, the obstinate pilot, along with Zarate.

This general of the Englishmen is . . . the same who, about five years ago, took the port of Nombre de Dios. He is . . . a man about 35 years of age, low of stature, with a fair beard, and is one of the greatest mariners that sails the seas. . . . His vessel is . . . a perfect sailer. She is manned with a hundred men, all . . . as practised [in war] as old soldiers from Italy. . . . He treats them with affection, and they treat him with respect. He carries with him nine or ten cavaliers [gentlemen]. . . . These form a part of his council which he calls together for even the most trivial matter, although he takes advice from no one.

The aforesaid gentlemen sit at his table, as well as a Portuguese pilot [da Silva]. . . . He is served on silver dishes with gold borders and gilded garlands, in which are his arms. He carries all possible dainties and perfumed waters. He said that many of these had been given him by the Queen. . . . He dines and sups to the music of viols. . . .

I managed to ascertain whether the General was well liked, and all said that they adored him.

RAID ON GUATULCO

On April 13, 1579, Drake entered the harbor of Guatulco, about three hundred miles south of Acapulco. He needed water, and he had apparently abandoned his plan to attack Acapulco. The Spanish garrison there, as described by the obliging Zarate, seemed too strong.

Guatulco, on the other hand, was hardly a major Pacific port, like Lima and Acapulco. It boasted one trading vessel at anchor, three or four houses, and a population of six Spaniards—three of whom escaped to the woods.

Drake's men looted the little settlement and its church and forced their three captives to guide them to water. Drake held a council on the *Golden Hind*, at which he announced his decision to return by way of the Moluccas. But he kept Nuño da Silva, his Portuguese pilot, away from this council; in fact, he gave da Silva the impression the *Golden Hind* would return by way of the Northwest Passage.

Then Drake released his prisoners and left da Silva on board the Spanish vessel in the harbor. After Drake sailed away, the Spaniards took da Silva and interrogated him. The small pilot in the long black coat, indignant with Drake for abandoning him, passed on the information that Drake would go back by the Northwest Passage—which is perhaps what Drake intended him to do.

Later, da Silva was bundled off to the Inquisition because he had attended Drake's Protestant services, and he was sentenced to perpetual exile from the Indies. Some scholars think that he eventually settled in Plymouth, England.

The following selection describes the raid on Guatulco; it is from a letter by Gaspar de Vargas, Alcalde (Mayor) of Guatulco, to the Viceroy of Mexico, Don Martin Enriquez (Nuttall, Zelia, ed., *New Light on Drake*).

This morning of Holy Monday at eight o'clock, being in the port of Guatulco, I was informed by some sailors of a ship belonging to Juan Madrid, which was in the said port . . . that they had just seen two sails very near to the entrance of port. One was large, the other small, and they inferred that the first was the vessel from Peru that they were expecting and that the smaller must be a bark used for pearl-fishing along this coast.

Two hours later, at about ten o'clock, both ships began to enter the port abreast, and it became apparent that the larger one was . . . of more than three hundred tons. The other one also appeared to be larger than had first been said.

They entered the port with great determination and the larger ship cast anchor. The bark, which turned out to be a launch, and the ship's boat, filled with men, began to come very suddenly, in a resolute manner, towards the shore. Then only was it understood that it was the English Corsair. . . .

I went to meet them on the shore with the few Spaniards and some Indians who were decorating the

Church for Holy Thursday and Easter. With the weapons that we found, we prepared to oppose their landing and succeeded in so far that the boat which carried more than forty archers and gunners was delayed until the launch began to discharge its artillery, which was supported by the arquebuses in the boat.

It then became necessary for us to abandon the town and retire up the [wooded] hill, from the heights of which we discharged our arquebuses. We saw them land, and with their captain, begin to plunder the property of the merchants and of those of us who live there.

What is most and above all else to be deplored is the shamelessness with which they, with their knives, hacked into pieces the sacred images and crucifixes, after which, laden with plunder, they returned to their ship. . . .

As far as we could see they carried off three persons, who were the curate; his relative the mayor of Suchitepec . . . who had come to spend Holy Week . . . and a certain Francisco Gomez, encomendero [holder of Indians].

At that time I went down and mustered a few Spaniards. When those of the ship saw this they returned to the shore in two boats so as to seize some of them. Not succeeding in doing this they returned to their ship.

At dusk I returned to the town for the third time to ascertain whether I could obtain some information as to who the men are. All that I have been able to find out is that the men on the ship belonging to Juan de Madrid think that the name of the pilot . . . is Morera. An Indian . . . recognised . . . one or two men who used to frequent this port as sailors.

I then came to this town of Guatulco and have just arrived here at ten o'clock at night, so as to send Your Excellency this dispatch by a suitable person. . . . From this same place I sent another Spaniard to San Juan Acapulco, a hundred leagues from here, so that even if he has to kill horses in doing so, he should reach that port before the ship [Drake's], so that the necessary precautions can be taken. . . .

After finishing this dispatch, I will return to the port to see what can be undertaken. I think that, beyond a doubt, he will take the route of Acapulco. . . . According to [the] sailors, the [Corsair's] ship is so low in the water that she appears . . . laden with gold, silver and merchandise. They therefore say and suspect that he must have done harm to . . . ships of Peru.

The viceroy, Don Martin Enriquez, made the following report about Drake's raid in a letter to King Philip II (Nuttall, Zelia, ed., *New Light on Drake*).

Today, Thursday, the twenty-third of April, after one o'clock of the afternoon, I received the enclosed letter. . . . It is from the Chief Alcalde of the port of Guatulco, and will inform Your Majesty of what is happening.

I cannot understand how [Drake and his men] came into this South Sea, if it is not by the same way that they came the last time, when they went to Panama. Now Your Majesty will see how important it is that this Sea should have security and will give orders to safeguard that Strait [of Magellan] which affords a troublesome entrance to this country, Peru and China [the Philippines].

Up to the present hour, I have no further light than that contained in the enclosed letter. I am now dispatching couriers to the inhabitants of the coasts so that they should be warned, although they can do nothing more than take refuge in the woods. For on those coasts there is no mode of defence. In many places there are not more than four Spaniards, excepting in the port of Acapulco, and even there there are very few Spaniards and not many Indians. . . .

I am also writing to the General Don Cristobal de Eraso [general of an armada in the West Indies], giving him an account of what is happening, so that, if possible, precautions can be taken against the time when the Englishmen return to the place whence they came, provided he has the men and is able to punish them.

Viceroy Don Martin Enriquez expressed further thoughts about Drake's raid in a letter to his fellow victim, the viceroy of Peru (Nuttall, Zelia, ed., *New Light on Drake*).

On the entire coast of this New Spain [Mexico] the corsairs cannot do any harm that would amount to much, because it is almost uninhabited by Spaniards as well as by Indians. Where there are Indians, there are very few and they live in straw huts.

There are a few Spanish settlements inland, and the trade from these ports to Nicaragua and Panama consists of some native wearing apparel, beans and things of little importance.

If the Strait could be fortified and manned, it would remedy everything. To attempt to keep fleets, as in the North Sea [the Atlantic], that would patrol this sea [the Pacific], would be a very troublesome and costly remedy.

A RELIGIOUS SERVICE

"You are not Christians but idolaters, who adore sticks and stones," Drake's boatswain told Gomez Rengifo, shipping agent of Guatulco, as he smashed the Spaniards' crucifix.

Drake's men broke the altar stone in the church, hacked the statues, wiped their faces on the altar cloths, and donned the priest's chasuble in mockery. Yet in typical Anglo-Saxon fashion, they combined this hatred of foreign practices with fair play for the beliefs of others by not forcing the Spaniards to attend their Protestant services on shipboard.

The following selection, from the "Deposition of Francisco Gomez Rengifo" (Nuttall, Zelia, ed., *New Light on Drake*), describes Drake conducting the evening worship in the harbor of Guatulco.

I saw how . . . the said Francis Drake had a table placed on deck at the poop of the vessel, and, at its head, on the floor, a small box and an embroidered cushion. He then sent for a book of the size of the *Lives of the Saints* and . . . struck the table twice with the palm of his hand.

Then, immediately nine Englishmen, with nine small books of the size of a breviary, joined him and seated themselves around him and the table. Then the said Francis Drake crossed his hands and, kneeling on the cushion and small box, lifted his eyes to heaven and remained in that attitude for about a quarter of an hour.

He then said to me and to the other prisoners that if we wanted to recite the psalms according to his mode we could stay, but if not, that we could go to the prow. As we stood up to go towards the prow . . . he began reading the psalms in the English language. . . . This act lasted about an hour and then they brought four viols, and made lamentations and sang together. . . .

Immediately afterwards he ordered a boy . . . to

come and then made him dance in the English fashion, with which the service ended. . . .

When he had finished . . . Francis Drake said to me "You will be saying now 'This man is a devil, who robs by day and prays at night in public.' This is what I do, but it is just as when King Philip gives a very large written paper to your Viceroy, Don Martin Enriquez, telling him what he is to do and how he is to govern, so the Queen, my Sovereign Lady, has ordered me to come to these parts. . . . If it is wrong it is she who knows best and I am not to be blamed. . . .

"But I do regret to possess myself of anything that does not belong exclusively to King Philip or to Don Martin Enriquez, for it grieves me that their vassals should be paying for them. But I am not going to stop until I have collected the two millions that my cousin John Hawkins lost . . . at San Juan de Ulua."

THE *GOLDEN HIND* DISAPPEARS

Don Martin Enriquez, viceroy of Mexico, was determined to catch "the English Corsair" this time. He rushed men to threatened ports like a ranger fighting a forest fire: two hundred to Acapulco, two hundred to Guatulco, two hundred to Guatemala, and two hundred to San Juan de Ulua on the Gulf of Mexico.

The two hundred men at Acapulco embarked in a hastily assembled fleet and sailed off to capture the pirates.

They were under just two handicaps. First, Drake had

left Guatulco over a month before. Second, they went south—and Drake was now above Acapulco, headed north. They were trying to catch him by running away from him.

The prefect of Mexico City, Don Luis de Velasco, was the only one to guess Drake's intentions. Don Luis gave Don Martin some excellent advice, which Don Martin did not take.

The following selection, from a letter of Don Luis de Velasco to King Philip II (Nuttall, Zelia, ed., *New Light on Drake*), states Velasco's theory of Drake's whereabouts.

When [the expedition prepared by the Viceroy, Don Martin Enriquez] set sail to pursue the Corsair, two months had passed since he had departed from . . . Guatulco . . . without their having had news of him from any part of this entire coast.

During all this time, unless he were forced to careen his vessel, he could have accomplished the dangerous part of his navigation, for he was provided with everything and must have had more desire to safeguard his rich prize than to seek ports in this New Spain and await vessels to fight with. . . .

[Also] it must be believed that he would not [repair his ship] on the coast by which he had come, doing damage, and where the people are irritated and in revolt against him. . . . The sending of the [Spanish] ships with orders to sail along the coast . . . from Acapulco [south] to the Realejo (which are three hundred leagues) was . . . purposeless, for even if the Corsair had . . . taken some port there, he would already have been sought out and chased by the ships from Guatemala and Lima, when these arrived. . . .

What they could do to some purpose would be to follow, without delay, the opposite direction, and go [north] towards the ports of Navidad, Colima, Costa de Chiametla, Culiacan and California. . . . The above would be the route he would have to follow on his search for the new strait [the Northwest Passage] of which they say he spoke so much.

Even if he did not find this strait, he would come across ports where he could stay and take shelter with great advantages of weather, as the country is very far north. . . . It would certainly be rash for him to endeavour to go by the Strait of Magellan . . . on account of the small number of men he carries. For it can be safely counted upon that the coasts, along which he would have to sail, are now prepared against him. . . .

As it does not appear to be out of the question that, if he does not find the strait he is looking for, he will follow the course of the Portuguese, I dared . . . to tell the Viceroy that it seemed certainly right to send a small vessel to the Philippines, with letters for Molucca, so that they should be warned of the coming of this Corsair and could be prepared to block his way in case he should arrive.

But my suggestion was to little purpose, for the Viceroy held a different opinion, and it seemed to him to be a mistake to send a ship for this purpose. It may be a mistake, but not a very great one, for as the route is a safe and certain one, he would run the risk of losing but little, and, possibly, gaining much. The result of this explanation has been that the Viceroy is displeased with me.

In a letter he wrote King Philip II (Nuttall, Zelia, ed., *New Light on Drake*), Don Martin Enriquez explained why he did not follow Don Luis de Velasco's advice.

Some persons here are expressing their belief that if the Corsair does not find another outlet he must perforce go by the Portuguese route and pass by regions where, if warning is given, he might be awaited. . . . It also appears to these persons that Molucca should be notified so that the Portuguese there, who are the Corsair's enemies, should know of his coming, particularly as he will be passing through the regions where they have their trade.

This may be, but I do not see that there is so much intercourse between those who are in China [the Philippines] and the Portuguese who are in Molucca. I would gladly notify the Governor of China [the Philippines], if this were possible, but on all this coast, for more than two hundred leagues, there are not more than three vessels: Your Majesty's, which is in the port of Acapulco; another small vessel which is laden with stuffs for Zonzonate [a town in modern El Salvador], and the vessel that was in Guatulco and the Corsair had in his power. . . .

And these two vessels, and others, if they were forthcoming, are more necessary for what I propose doing, which is to send them to look for the Corsair, than to send them to China [the Philippines], even if the weather permitted this.

Moreover, I do not believe that the Corsair will go there, nor is it a region where any damage could be done on land, for, after having left his country with two hundred and sixty men, his galleon now only carries eighty. God knows how many will arrive over there.

THE NORTHWEST PASSAGE

The Northwest Passage, a kind of misplaced Panama Canal thought to flow through Canada, existed only in the imagination of geographers. Yet it was given a name, "The Strait of Anian," and it appeared on many sixteenth-century maps. It would have made an ideal escape route for Drake—a remote blue waterway, out of sailing range of the Spaniards. But whether Drake planned to look for this passage when he left Plymouth or whether he actually sought it now are disputed questions.

All that is known is that when Drake left Guatulco on April 16, 1579, he made a great loop out into the Pacific to avoid opposing winds. When he reached the high latitudes of the northern Pacific, at a point 200 miles west of the coast, he picked up the prevailing northwest gale which swept his two ships—the *Golden Hind* and the captured Spanish bark—back to the coast of California.

Here, far north of San Francisco, Drake did not have to be looking constantly over his shoulder for Spanish pursuers. He may have been as far north as Vancouver, but the best evidence seems to indicate that he reached California near Cape Mendocino. His men, accustomed to the tropics, were shivering in the chill westerlies and thick fogs, so he turned south.

As Drake voyaged south, it became warmer—but he found no Strait of Anian, and he decided there was none. Just above San Francisco Bay, he established a base camp. It was probably on the inlet, shaped like an upside-down bowl, behind Point Reyes—now called Drake's Bay. Here Drake could accomplish the two objectives of his California detour: to overhaul and clean the *Golden Hind*; and to delay his departure for the Moluccas so as to arrive there when the typhoon season was past.

The following selection, from *The World Encompassed* (1628), describes Drake's voyage to California. Its statement that he reached 48° north latitude has, however, been questioned.

From *Guatulco* we departed the day following, viz., *April* 16 [1579], setting our course directly into the sea, whereon we sailed 500 leagues in longitude, to get a wind and between that and *June* 3, 1400 leagues in all, till we came into 42 deg. of North latitude, where in the night following we found such alteration of heat, into extreme and nipping cold, that our men . . . did grievously complain thereof. . . .

The 5 day of *June*, we were forced by contrary winds to run in with the shore, which we then first descried, and to cast anchor in a bad bay . . . where we were not without some danger by reason of the many extreme gusts . . . which . . . were . . . followed [by] . . . thick and stinking fogs, against which the sea prevailed nothing, till the gusts of wind again removed them. . . .

From the height [latitude] of 48 deg., in which now we were, to 38, we found the land, by coasting alongst it, to be but low and reasonably plain. . . . In 38 deg. 30 min. we fell with a convenient and fit harbor, and *June* 17 came to anchor therein, where we continued till the 23 day of *July* following. . . .

We conjecture, that either there is no passage at all [to the Atlantic] through these Northern coasts (which is most likely), or if there be, that yet it is unnavigable. Add hereunto, that though we searched the coast diligently, even unto the 48 deg., yet found we not the land to trend so much as one point in any place towards the East, but rather running on continually North-west as if it went directly to meet with Asia.

And even in that height [latitude], when we had a frank wind to have carried us through had there been a passage, yet we had a smooth and calm sea, with ordinary flowing and reflowing, which could not have been had there been a frete [strait]; of which we . . . infallibly concluded . . . that there was none.

INDIANS OF SAN FRANCISCO BAY

Friendly Indians now visited Drake's fortified camp. They had a primitive culture, but they smiled and exchanged presents with the tourists from Plymouth.

Drake's men were sure their visitors admired them. It was an article of faith that Indians hated cruel Spaniards but loved bold, brave Englishmen. When, a week after their arrival, the chief placed a "crown" of feathers on Drake's head, Drake quick-wittedly accepted it—and the entire territory of the Indians—as a present to Queen Elizabeth! It is extremely unlikely, however, that the chief thought he was giving away his kingdom.

The following selection, from *The World Encompassed* (1628), describes the customs and houses of these California Indians as well as the gift-giving.

The next day after our coming to anchor . . . the people of the country shewed themselves, sending off a man . . . to us in a canoe. Who . . . spake to us continually as he came rowing on. And at last . . . staying [stopping] himself, he began more solemnly a long and tedious oration. . . . And after-[wards] . . . returned back to shore again.

He shortly came again . . . and so the third time, when he brought with him (as a present from the rest) a bunch of feathers, much like the feathers of a black crow, very neatly . . . gathered upon a string. . . . With this also he brought a little basket made of rushes, and filled with an herb which they called *Tabah*. Both which being tied to a short rod, he cast into our boat.

Indians in California crowned Drake, who named his new kingdom Nova Albion and claimed it for Elizabeth I. Drake had a brass plate made stating these facts, which is being fastened to a post in the background.

Our General intended to have recompensed him immediately with many good things . . . but . . . he could not be drawn to receive them . . . save one hat, which being cast into the water out of the ship, he took up. . . .

The 3 day following, viz., the 21 [of June, 1579], our ship having received a leak at sea, was brought to anchor nearer the shore, that, her goods being landed, she might be repaired; but . . . our General first of all landed his men . . . to build tents and make a fort for the defence of our selves and goods. . . .

Which when the people of the country perceived us doing . . . in great haste . . . with such weapons as they had, they came down unto us, and yet with no hostile . . . intent . . . standing, when they drew near, as men ravished in their minds . . . their errand being . . . . to worship us as Gods.

Drake had shirts and linen cloth given to these Indians; in return, the Indians gave the Englishmen feathers, quivers, and skins of beasts.

Having thus had their fill of . . . visiting and beholding . . . us, they departed with joy to their houses, which houses are dug round within the earth, and have from the uppermost brims of the circle clefts of wood set up, and joined close together at the top, like our spires on the steeple of a Church; which being covered with earth, suffer no water to enter, and are very warm.

The door in . . . most . . . of them performs the office also of a chimney to let out the smoke. . . . Their beds are the hard ground, only with rushes strewed upon it, and lying round about the house, have their fire in the midst, which by reason that the house is but low vaulted, round, and close, giveth a marvelous reflexion to their bodies to heat the same.

Their men for the most part go naked. The women take a kind of bulrushes, and . . . make themselves thereof a loose garment, which . . . hangs down about their hips . . . ; about their shoulders they wear also the skin of a deer. . . . They are very obedient to their husbands. . . .

As soon as they . . . returned to their houses, they began . . . a kind of most lamentable weeping and crying out; which they continued a great while together. . . .

Against the end of three days more . . . were assembled the greatest number of people. . . . Amongst the rest the king himself, a man of a goodly stature and comely personage, attended with his guard of about 100 tall and warlike men, this day, viz. *June* 26, came down to see us. . . .

In their coming forward they cried continually after a singing manner, with a lusty courage. . . . In the forefront came a man of a large body and goodly aspect, bearing the Scepter or royal mace, made of . . . black wood . . . before the king.

Whereupon hanged two crowns, a bigger and a less, with three chains. . . .

The crowns were made of knitwork . . . with feathers of divers colours. . . . The chains seemed of a bony substance, every link . . . being very little, thin, most finely burnished. . . .

Next unto him that bore this Scepter, was the king himself, with his guard about him. . . . Upon his head was a caul of knitwork . . . somewhat like the crowns . . . ; upon his shoulders he had . . . a coat of the skins of conies [rabbits—actually ground squirrels], reaching to his waist. His guard also had each coats of the same shape, but of other skins. . . . After these . . . did follow the naked sort of common people. . . . Every one had his face painted, some with white, some black, and some with other colours. . . .

And being now come . . . near our fort, the Scepter bearer . . . began a song, and . . . a dance; whom the king with his guard and every other . . . person following, did in like manner sing and dance, saving only the women, who danced but kept silence. . . .

After . . . they had . . . tired themselves in this manner, they made signs to our General to . . . sit down; unto whom both the king and divers others made several orations, or . . . supplications, that he would take the Province and kingdom into his hand, and become their king. . . . [Then] the king himself, with all the rest . . . joyfully singing a song, set the crown upon [Drake's] head. . . .

These things being so freely offered, our General thought not meet [proper] to . . . refuse the same, both [because] . . . he would not give them any cause of . . . disliking . . . him . . . and . . . [because] he knew not . . . what honour and profit it might bring to our country in time to come.

Wherefore, in the name . . . of her most excellent majesty, he took the scepter, crown, and dignity of the said country into his hand.

NOVA ALBION

Drake named his newly-acquired kingdom *Nova Albion* –Latin for New England. The Puritans should have asked Drake's permission when they called the opposite side of the continent New England half a century later!

Albion, perhaps from a Celtic word, is the ancient name for England. Scholars in Drake's day, however, also connected the word with the Latin *albus*, meaning white, and they thought England had been named for the white cliffs of Dover. Drake too may have been influenced in his choice of the name by the white cliffs and white sand dunes he found along the California coast.

Some have thought that Drake planned to return to California later and establish a permanent colony, but there is little evidence for their theory. More likely he had in mind simply a supply base for English vessels sailing to the Moluccas by way of the Northwest Passage–if it was ever found. He thought the Spaniards had not come this far north, but, in fact, Cabrillo had been here in 1542.

Nothing came of *Nova Albion.* Yet its appearance on sixteenth-century maps after Drake's voyage is a harbinger of the future. The mighty empire which one man, Francis Drake, had the audacity to challenge was to prove weaker than Europe could have supposed. Drake had come to raid and run, but soon his countrymen would come to stay, despite Spain's claims to ownership. In the end, half the New World would fall to Englishmen, and it would become the cradle of a liberty wider even than that of England.

The following selection, from *The World Encompassed* (1628), briefly describes *Nova Albion,* that is, northern California.

After . . . our necessary businesses were well dispatched, our General, with his gentlemen and many of his company, made a journey up into the land . . . to be the better acquainted with the nature and commodities of the country.

The inland we found to be far different from the shore, a goodly country, and fruitful soil. . . . Infinite was the company of very large and fat Deer which . . . we saw . . . besides a multitude of a strange kind of Conies [rabbits—actually ground squirrels]. . . . Their heads and bodies . . . are but small; his tail, like the tail of a Rat, exceeding long; and his feet like the paws of a . . . mole; under his chin, on either side, he hath a bag, into which he gathereth his meat . . . that he may with it, either feed his young, or feed himself when he lists not to travel from his burrow. The people eat their bodies, and make great account of their skins, for their king's holiday's coat was made of them.

This country our General named *Albion*, and that for two causes; the one in respect of the white banks and cliffs, which lie toward the sea; the other, that it might have some affinity, even in name also, with our own country, which was sometime so called.

Before we went from thence, our General caused to be set up a monument of our being there, as also of her majesty's and successor's right and title to that kingdom; namely, a plate of brass, fast nailed to a great and firm post.

[Thereon] is engraven her grace's name, and the day and year of our arrival there, and of the free giving up of the province and kingdom, both by the king and people, into her majesty's hands: together with her highness' picture and arms, in a piece of sixpence . . . shewing itself by a hole made of purpose through the plate. Underneath was likewise engraven the name of our General, etc.

The Spaniards never had any dealing, or so much as set a foot in this country, the utmost of their discoveries reaching only to many degrees Southward of this place.

An ancient brass plate was found near Drake's Bay, just above San Francisco, in 1937; it bears the following inscription:

The brass plate found in 1937 near Drake's Bay, just above San Francisco, California.

BEE IT KNOWNE VNTO ALL MEN BY THESE PRESENTS IVNE 17 1579
BY THE GRACE OF GOD AND IN THE NAME OF HERR
MAIESTY QVEEN ELIZABETH OF ENGLAND AND HERR
SVCCESSORS FOREVER I TAKE POSSESSION OF THIS
KINGDOME, WHOSE KING AND PEOPLE FREELY RESIGNE
THEIR RIGHT AND TITLE IN THE WHOLE LAND VNTO HERR
MAIESTIES KEEPEING NOW NAMED BY ME AN TO BEE
KNOWNE VNTO ALL MEN AS NOVA ALBION.

FRANCIS DRAKE

Some historians accept this as Drake's plate, but others doubt its authenticity.

Drake was now almost ready to depart. The dark hull of the *Golden Hind* was newly-calked, washed, and shining. Large supplies of food and water had been taken on. The crew were rested.

But the longest, most dangerous part of the navigation lay ahead—into unknown seas, through the Indonesian labyrinth of shoals and islands, and around Africa's stormy Cape of Good Hope. Even with Colchero's charts and the big Portuguese map, Drake would have to continue wooing Lady Luck if he hoped to deliver his gold and silver in London.

Chapter Seven

THE WORLD ENCOMPASSED

On the bronze cannon which [Drake's] launch carried there was engraved the round world with a north star above it. . . . These were his arms which the Queen had given him, commanding him to go around the world.

—NUÑO DA SILVA

Seven

ACROSS THE PACIFIC

On July 23, 1579, accompanied by doleful cries from the Indians who lit fires on the hilltops, the *Golden Hind* weighed anchor in the northern California bay. The captured Spanish bark was abandoned there. Along with the *Elizabeth* and *Marigold*, Drake had lost about half of his original 140 to 164 men, and he could man only one vessel for the long voyage ahead.

The next day, July 24, Drake stopped at an island group he named St. James, and the sailors killed seals and birds to stock the ship with more food. Then, for over two months, Drake navigated through the Pacific without sighting land. On September 30, 1579, he arrived at one of the Pelew Islands, on the doorstep of the Philippines.

Drake was fortunate to avoid storms or a shortage of provisions; Magellan's men, on his crossing, had had to eat the leather wrappings on the masts, and many had died from scurvy. Drake's course was shaped like a chaise longue—that is, slanting down southwest from San Francisco to a point just south of the Hawaiian Islands in 10° latitude, then straight west for five thousand miles to the Pelews and the Philippines. From the Philippines he would drop down, due south, to the "floor," the Moluccas.

At an island in the Pelews, Drake obtained fresh food

and water, but he was almost overrun by the Polynesians who brought supplies. These tall strong natives skimmed out to the *Golden Hind* in outrigger canoes. They were handsome but indolent. When they shifted from trading to thieving, Drake faced an annoyance and some danger.

The following selection, from *The World Encompassed* (1628), describes Drake's voyage and his first encounter with men of the East.

The 23 of *July* [the Indians] took a sorrowful farewell of us, but . . . ran to the top of the hills to keep us in their sight as long as they could, making fires . . . and . . . burning therein . . . sacrifices at our departure.

Not far without this harbor did lie certain Islands (we called them the Islands of Saint *James*), having on them plentiful and great store of Seals and birds . . . whereon we found . . . provision. . . . We departed again . . . *July* 25.

And our General [Drake] now . . . with general consent of all, bent his course directly to . . . the Islands of the Moluccas. And so having nothing in our view but air and sea, without sight of any land for . . . 68 days . . . we continued our course through the main Ocean, till *September* 30 . . . [when] we fell in ken of certain Islands, lying about eight degrees to the Northward of the line [the equator].

From these Islands . . . came a great number of

canoes, having in each . . . four . . . six . . . fourteen or fifteen men, bringing with them coconuts, fish, potatoes, and certain fruits. . . .

Their canoes were made . . . of one tree, hollowed within with great art . . . being made so smooth . . . that they bore a gloss. . . . A prow and stern they had of one fashion [alike], yielding inward in . . . a semicircle . . . and hanged full of . . . white . . . shells. . . . On each side of their canoes lay . . . two pieces of timber [i.e., outriggers]. . . at the end whereof was fastened crosswise a great cane . . . to keep their canoes from overthrowing. . . .

The first . . . canoes being come to our ship . . . began in peace to traffic [trade] with us, giving us one thing for another . . . [and] entreating us by signs most earnestly to draw nearer towards the shore. . . .

But these passing away and others continually resorting, we were quickly able to guess at . . . what they were; for if they received anything . . . they would neither give recompense nor restitution of it, but thought whatever they could finger to be their own, expecting always with brows of brass to receive more. . . . Yea, being rejected for their bad dealing . . . they . . . attempt[ed] to revenge themselves . . . and having stones . . . in their canoes, let fly . . . against us. . . .

That they might know that he had power to do

them harm . . . he [Drake] caused a great piece [gun] to be shot off, not to hurt them, but to affright them. Which wrought the desired effect . . . for at the noise . . . every one leaped out of his canoe into the water, and diving under the keel of their boats, stayed [kept] them from going any way till our ship was gone a good way from them. Then they all lightly recovered into their canoes, and got them with speed toward the shore.

Notwithstanding, other new companies . . . continually made resort to us . . . and . . . fell a filching of what they could. And one of them pulled a dagger and knives from one of our men's girdles, and being required to restore it again, he rather [tried] . . . to catch at more.

Neither could we at all be . . . rid of this ungracious company, till we made some of them feel some smart as well as terror; and so we left that place . . . to be known hereafter by the name of the *Island of Thieves.*

The "smart" which Drake inflicted upon his shop-lifting visitors was killing twenty of them. The name "Island of Thieves" was certainly borrowed from Magellan, who had a similar experience at Guam and gave it and its neighbor islands the same name (in Spanish, *Islas de los Ladrones*). Drake carried an account of Magellan's voyage with him.

A PORTUGUESE GALLEON

From the Island of Thieves Drake continued west until he bumped into Mindanao in the Philippines, the dead end of his course. Then he turned south.

In October he entered the Celebes Sea, which is fringed with islands and dangerous shoals, and looked for a pilot. Drake was a superb ocean navigator, but for local waters he usually planned to kidnap a local pilot.

Before he could pick up a native however, Drake encountered a large galleon manned by Portuguese. He was in seas ruled for three-quarters of a century by these master mariners, and it behooved him to be cautious. But Drake could never pass up a prize. That night he sailed close behind his intended prey—but morning brought a surprise.

The following selection narrates Drake's journey southward and his first meeting with the Portuguese in the East. It is from the report of Francisco de Dueñas, a Spanish spy in the Moluccas at this time (Wagner, Henry R., *Sir Francis Drake's Voyage Around the World*).

A Portuguese galleon on her way from Malacca to the Moluccas, when between the Island of Mindanao and Celebes sighted a ship with topmasts, and thinking it was Spanish because near the Philippines . . . sent . . . a boat . . . to offer assistance, believing that the other ship was out of her course.

Having arrived near they [the Portuguese in the boat] asked for the captain but no one answered as only two men were seen aboard. These made signs . . . to keep off and they [the Portuguese] returned half frightened to their ship and told the Captain what had occurred. As it seemed to him to be a small ship he gave no more thought to the matter, continuing on his way.

Night coming on, the Englishman followed [the Portuguese] ship, spending part of the night, as it appeared, in getting ready his guns. When day dawned, the ships were only half a league apart, the English vessel coming towards the galleon with pennants and flag flying.

When near enough to be heard, an Englishman spoke, saying: "Captain don Francisco . . . Englishman and Lutheran, orders you to strike your sails and surrender, and if you do not do so at once he will make you do so by force."

The Portuguese marvelling to hear what the Englishman had said, answered that they should come aboard, beginning at the same time to open some port-holes and get ready some guns which they carried.

The Englishman commenced to fire at them, but as [the Portuguese galleon] seemed like a large ship with many men aboard, after firing seven or eight shots he kept at a distance. . . .

Keeping at a distance, as I said, the [English] ship took a southeast course, which is the one for the Moluccas, and in a short time was lost to sight.

Another day the Englishman arrived at Siago [the island of Siau] where he took two Indian fishermen, who thinking they were Portuguese, had gone to them. By signs they [the English] asked them to show them the road to the Moluccas and the Indians said they would.

So these [Indians] guided them and when near the Moluccas a Portuguese came along with a small vessel looking for supplies for those islands, and thinking it was a Portuguese ship coming from Malacca, went straight to her and going aboard was very frightened to see such strange people.

The English, however, treated him well, entertaining him with kind words and telling him not to fear, that they were Englishmen and Lutherans, but did no harm to any one. They asked him where the fortress of the Portuguese was and he said very near.

Among other questions, they asked him what direction the Portuguese took when they went to Malacca, placing before him a sailing chart. He answered that he had been born and brought up in Ternate, which was the truth, and that he knew nothing about navigation or sailing charts. Not pursuing the subject further, they discussed other matters.

At this time the ship was seen from Ternate and the King sent at once two *caracoas* [large open boats] to find out what ship it was and from whence it came. When these [the men of Ternate] found that they were neither Portuguese nor Spaniards but English Lutherans . . . they returned to tell the King and he soon sent to invite them to come ashore at once, offering them port room and everything necessary, and saying that they should not go where the Portuguese were as they [the Portuguese] had a galley and a galleon with which they could do much damage.

THE SPICE ISLANDS

The Portuguese galleon was too heavily armed to be seized, so Drake sailed on his way. On November 3, 1579, he sighted the first of "four high peaked Islands" which make a dotted line running north to south, ending with a fifth, much larger one. These are the islands of Ternate, Tidore, Mortier, Makian, and Bachian.

These islands, crowned with volcanoes, were the world-famous Moluccas. Here grew ginger, nutmeg, and above all the spicy cloves so much in demand in western Europe for seasoning a monotonous winter diet. Harvested from thirty-foot evergreens, the spike-shaped crimson cloves (which turn black later) increased in value a hundredfold between Indonesia and London. If Drake had not captured the *Cacafuego*, his eyes would have sparkled at this prospect for gain.

He did not overlook the opportunity anyhow. Piloted by Lady Luck to Ternate—not to Tidore, the Portuguese stronghold—he found a "Cold War" going on between Baber, the aggressive Moslem ruler, and the Portuguese. Baber had recently driven the Portuguese from Ternate and was now looking for a new customer for his cloves.

Baber promptly offered to sell his cloves only to the English and to allow them to erect "factories" (warehouses and trading posts) on Ternate in return for Drake's promise to send out future trading expeditions. When members of the East India Company first arrived in the East in 1600, they based their claim to trading rights on the verbal agreement reached by Drake and Baber.

On the surface, all was Oriental courtesy and complaisance, music and good fellowship. But beneath the surface there was a dangerous tension. Drake, good-humored but highhanded, and the wily, powerful Sultan were two of a kind. Drake attempted to load the cloves without paying the Sultan's 10 per cent export tax; Drake thought his cannon canceled the tax. Thereupon Baber decided to cancel Drake and ordered him put to death—presumably by a treacherous attack. This is what lies behind the suspicions of Drake's men, related in the excerpt that follows, and their refusal to allow Drake to go ashore.

In the end, Drake gave Baber some luxurious presents, and the trade and treaty went through. According to one source, Drake took on six tons of cloves, of which three tons may have reached London. But Drake also decided that five days in this languorous isle were plenty, and he prepared to depart.

The following selection, from *The World Encompassed* (1628), describes Drake's arrival and the ceremonies on Ternate.

The third of *November* we came in sight of the Islands of the *Moluccas*, as we desired. . . .

We directed our course to have gone to *Tidore*, but in coasting along a little Island belonging to the king of *Ternate*, *November* 4, his deputy or Viceroy . . . came off to our ship in a canoe. . . . Who . . . entreated our General . . . to run with *Ternate*, not with *Tidore*, assuring him that his king would be wondrous glad of his coming. . . .

On these persuasions our General resolved to run

with *Ternate*, where the next day, very early in the morning, we came to anchor: and presently our General sent a messenger to the king. . . .

[Meanwhile] the king . . . [had been] so . . . moved to the well liking of the matter [by the report of his Viceroy], that before our messenger could come half the way, he had sent the Viceroy [back] . . . to our General, with special message that he [Drake] should not only have what things he needed . . . but that . . . he would sequester the commodities and traffic [trade, especially in spices] of his whole Island from others, especially from his enemies, the Portuguese . . . and reserve it to . . . our Nation. . . . In token whereof he had now sent to our General his signet, and would within short time after come in his own person. . . .

The manner of his coming [a little later] was princely. . . . First . . . did he send off 3 . . . large Canoes [or *caracoas*], in each whereof were certain of the greatest personages . . . attired . . . in white Lawn . . . having over their heads, from one end of the Canoe to the other, a covering of thin and fine mats, borne up by a frame of reeds, under which every man sat. . . . Besides these were . . . young and comely men, a great number attired in white . . . under the same covering, but in inferior order. . . .

The rest of the men were soldiers, who stood . . . round about on both sides; on the outside of whom, again did sit the rowers. . . . In the forepart of each

Canoe, sat two men, the one holding a Tabret, the other a piece of brass, whereon they both at once struck; and observing a due time . . . between each stroke . . . directed the rowers. . . . Each [canoe had] at least one small cast piece [cannon]. . . . Besides every man except the rowers, had his sword, dagger, and target [shield], and some of them other weapons, as lances, calivers, bows, arrows, and many darts. . . .

The king himself was not far behind . . . with 6 grave and ancient fathers in his Canoe [or *caracoa*]. . . . He was of a tall stature, very corpulent and well set together. . . . Neither his Viceroy . . . nor any other of his counsellors, durst speak unto him but upon their knees. . . .

He was received in the best manner we could. . . . Our ordnance thundered, which we mixed with . . . small shot, among which sounding our trumpets. . . . Wherewith he was so much delighted, that requesting our music to come unto the boat [*i.e.*, asking the English musicians to enter the small boat towed by the *Golden Hind*] he joined his Canoe [or *caracoa*] to the same [boat], and was towed at least a whole hour together, with the boat at the stern of our ship. Besides this, our General sent him . . . presents. . . .

The king, as soon as we were come to anchor, craved pardon to be gone, and so took leave, promising us that the next day he would come aboard. . . .

At the time appointed, our General . . . looked

for the king's return, who failing [to come] . . . sent his brother to make his excuse, and to entreat our General to come on shore. . . . Our General could willingly have consented, if the king himself had not first broke his word. . . . [Because of this] the whole company . . . by no means would give consent he [Drake] should hazard himself. . . . Our General [therefore] . . . sent certain of his gentlemen to the Court. . . .

Drake's messengers were taken to the "council-house," a kind of open pavilion, at one side of which was "the chair of state, having directly over it . . . a . . . rich canopy."

The king at last coming . . . with 8 or 10 . . . Senators following him, had a very rich canopy . . . borne over him, and was guarded with 12 lances, the points turned downward. Our men . . . arose to meet him, and he very graciously did welcome and entertain them. . . .

His attire . . . from the waist to the ground was all cloth of gold . . . his legs bare, but on his feet a pair of shoes of cordovan, dyed red. . . . [On] his head were . . . rings of plated gold . . . resembling a crown. . . . About his neck he had a chain of perfect gold. . . . On his left hand was a Diamond, an Emerald, a Ruby, and a Turquoise . . . on his right hand, in one ring, a big and perfect Turquoise and in another ring many Diamonds. . . .

He sat in his chair of state, at his right side . . . a page with a . . . fan (richly embroidered . . . with Sapphires) . . . gathering the air to refresh the king, the place being very hot. . . . Our gentlemen . . . delivered their message, and received answer. [Then they] were licensed to depart, and were safely conducted back [to the ship]. . . .

By the ninth of *November*, having gotten what provision the place could afford us, [and as much cloves as we desired], we then set sail.

REFITTING—FAREWELL TO MARIA

By now the *Golden Hind* and its crew were both again in need of repairs. The ship's bottom was foul with barnacles; the seamen were exhausted or sick.

So Drake sailed for only five days, southwest, until he came to a heavily forested islet off the coast of Celebes. Celebes is a large octopus-shaped island lying west and southwest of the stepping-stone Moluccas.

Here Drake had his carpenter build stocks for the ship. An entrenchment was dug around the stocks, and then the men unloaded the *Golden Hind* and drew it up on the stocks, where they could get at its bottom. When not chipping away the barnacles, they could swim, stretch out in the sun, or chase crayfish that scuttled under the roots of huge trees and even climbed into the branches. The sailors named their refuge Crab Island after the crayfish.

At night they slept in tents behind their dirt ramparts. The piles of silver bars gleamed in the moonlight

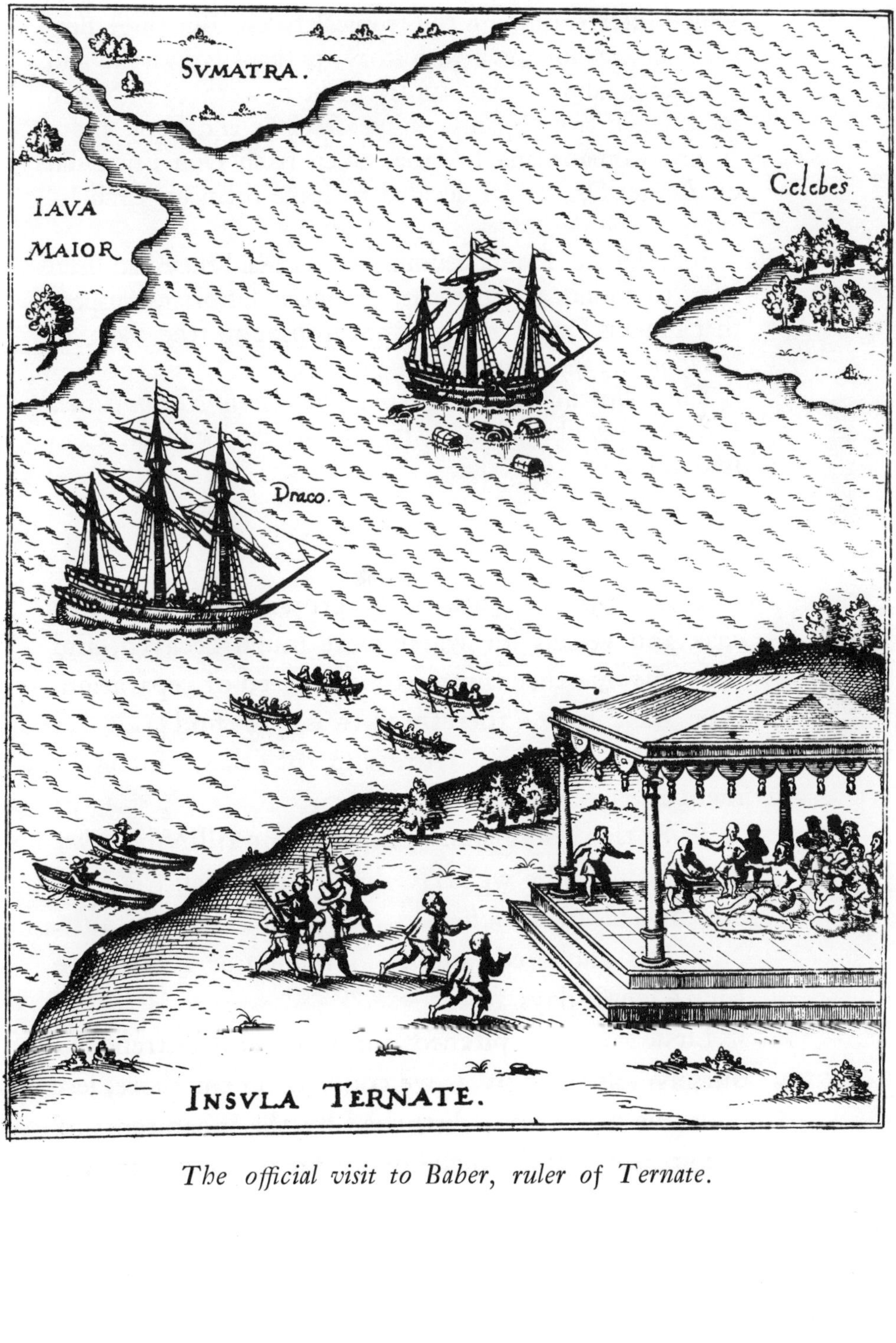

The official visit to Baber, ruler of Ternate.

which filtered through the tropical forest. But three persons, almost unknown to history, perhaps lay awake and wondered about their future. These were two Negro men, taken from Spanish ships off Chile and Peru, and a Negro woman named Maria, taken from Don Francisco de Zarate. Drake had decided to leave them on Crab Island.

John Drake, Francis's young cousin, said they had been left "to found a settlement"; modern historians suspect that Francis Drake did not want to have to feed them on the long voyage ahead.

The following selection, from *The World Encompassed* (1628), describes Drake's twenty-six day stay on Crab Island.

Our ship . . . was now grown foul . . . our cask and vessels for water were much decayed, and . . . divers other things stood in need of reparation. Our next care [therefore] was [to find] . . . a place . . . for the redressing of these inconveniences.

With this resolution we sailed along till *November* 14, at what time we arrived at a little Island (to the Southward of *Celebes*) . . . without inhabitants. . . . We anchored, and . . . made our abode here for 26 whole days. . . .

First . . . we pitched our tents and entrenched ourselves as strongly as we could upon the shore, lest . . . we . . . [be] disturbed by the inhabitants of the greater Island which lay not far to the Westward.

. . . [Then] we landed our goods, and had a Smith's [blacksmith's] forge set up, both for the making of some necessary shipwork, and for the repairing of some iron-hooped casks. . . . And [since] our Smith's coals were all spent . . . there was order given . . . for the burning of charcoal. . . .

We trimmed our ship, and performed our other businesses. The place afford[ed] us . . . wonderful refreshing to our wearied bodies. . . . We in short space grew . . . to be strong, lusty, and healthful. . . .

The whole Island is a . . . wood, the trees . . . large and high . . . without boughs, save only in the very top. . . . Among these trees, night by night, did shew themselves an infinite swarm of fiery-seeming-worms flying in the air . . . as if every twig on every tree had been a lighted candle. . . . Here we saw [also] . . . a multitude of huge Bats. . . . They fly with marvelous swiftness, but their flight is very short; and when they light, they hang . . . by the boughs with their backs downward. . . .

[We saw also a] multitude of . . . Crayfish, of such a size, that one was sufficient to satisfy four hungry men at a dinner. . . . They are . . . utter strangers to the sea, living always on the land, where . . . they dig great . . . caves under the roots of . . . huge . . . trees. . . . Some . . . when we came to take them, did climb up into the trees to hide . . . whither we were enforced to climb after them. . . . This Island we called *Crab-Island*.

According to John Drake, Francis Drake's cousin, in his "First Declaration" made before Alonso de Vera in Paraguay, it was on Crab Island that Drake left the two Negro men and the Negro woman Maria to found a settlement, leaving them rice, seeds, and means of making fire.

All necessary causes of our staying longer in this place being at last finished . . . and having the day before furnished our selves with fresh water from [a nearby] Island, and taken in provision of wood . . . *December* 12 we put to sea, directing our course toward the West.

A REEF IN THE NIGHT

When Drake left Crab Island, December 12, 1579, the northeast monsoon had just begun to blow. With this at his back, Drake planned to sail south to the line of islands —Sumatra, Java, Timor, and others—which are like a saucer holding the Indonesian archipelago. Beyond this saucer lies the Indian Ocean, which Drake was eager to reach.

For several weeks Drake maneuvered through the numerous shoals and islets off the eastern coast of Celebes. At last, on January 9, 1580, he came to a point where the coast curved west. The *Golden Hind* must have reached the bottom of the Celebes. Now there would be only a short stretch of open sea between Drake and the final barrier of islands which included Java and Timor.

Drake breathed a sigh of relief. The worst was over. That evening he added more sail, and the *Golden Hind* flew before a gale, under the southern stars. The psalms having been sung, the seamen rested or dreamed of the hills of Devon.

Suddenly, at 8 P.M., there was a hard jolt. Sleeping sailors were jarred awake. Those standing on watch staggered. The ship stopped.

Twelve thousand miles from home, the *Golden Hind* had struck a shelving reef and was hard aground. The rock was on the port side; the monsoon, blowing steadily from the starboard, held the ship on the reef and began to force it higher.

Thump–thump–thump—white-faced, the crew listened as the waves lifted the hull, then dropped it against the sharp rock. Soon the seams would open to the sea.

The following selection, from *The World Encompassed* (1628), describes the greatest peril of Drake's voyage.

The 16 day we had sight of the Island *Celebes*, but having a bad wind, and being entangled among many Islands . . . we could not by any means . . . continue on our course farther West, but were enforced to alter the same toward the South; finding that course also to be both difficult and very dangerous by reason of many shoals . . . among the Islands. . . .

Thus were we forced to beat up and down with extraordinary care . . . till *January* 9, at which time we supposed that we had at last attained a free passage . . . and the wind . . . followed us as we desired with a . . . gale.

When lo, on a sudden, when we least suspected . . . and we were now sailing onward with full sails, in the beginning of the first watch . . . at night, even in a moment, our ship was laid up fast upon a desperate shoal. . . .

The more we looked the less hope we had of getting clear of it. . . . We . . . humbly besought Almighty God to extend his mercy unto us . . . and so preparing as it were our necks unto the block, we every minute expected the final stroke. . . .

Yet . . . as soon as prayers were ended, our General . . . encouraged us all to bestir our selves, shewing us the way thereto by his own example; and first of all the pump being well plied, and the ship freed of water, we found our leaks to be nothing increased, which . . . gave us some hope of respite. . . .

Our next assay was for . . . anchorhold to seaward of us (whereon to haul) . . . to clear ourselves. . . . But even a boat's length from the ship, our General found that the bottom could not by any length of line be reached . . . so that the beginnings of hope . . . were . . . quite dashed again. . . . [But] our General . . . dissembl[ed] . . . and [gave] . . . cheerful speeches and . . . encouragement. . . .

The day . . . at length appearing, and it being almost full sea . . . we again renewed our travail to see if we could now possibly find any anchor hold. . . . But this second attempt proved as fruitless as the former, and left us nothing to trust to but prayers. . . . And that our faith might be the better strengthened . . . we had a Sermon, and the Sacrament of the body and blood of our Saviour celebrated.

After this . . . we fell to . . . unloading of our ship by casting some of her goods into the sea . . . but [could not free ourselves]. . . . The place whereon we sat so fast was a firm rock in a cleft, whereof . . . we stuck on the larboard side. At low water there was not above six foot depth in all on the starboard [but] with little distance as you have heard no bottom to be found. The breeze during the whole time . . . blew somewhat stiff directly against our broadside, and so perforce kept the ship upright.

It pleased God in the beginning of the tide, while the water was yet almost at lowest, to slack the stiffness of the wind; and now our ship, who required thirteen foot water to make her float, and had not at that time on the one side above seven at most, wanting [lacking] her prop [the wind] on the other side . . . fell a heeling towards the deep water, and by that means freed her keel and made us glad men. . . .

Of all the dangers that in our whole voyage we met with, this was the greatest; but it was not the last, as may appear by what ensueth.

THE CHAPLAIN IS EXCOMMUNICATED

While on the reef, Drake had tried everything. He threw overboard three tons of cloves, two cannon, munitions, and even some food supplies. But it was only the startling shift of wind at 4 P.M. the next day that saved the *Golden Hind.*

For a short time the northeast monsoon ceased, and the wind blew from the south. Drake immediately ordered all sails hoisted, and the *Golden Hind* slid off the reef into deep water.

Among the things Drake had tried was prayer. Chaplain Fletcher, articulate and still resentful of the execution of gentleman Doughty by commoner Francis Drake, preached a sermon.

"The judgment of God!" thundered Fletcher. The wreck off Celebes was God's punishment for what they had done in executing Doughty at the "Bloody island" in the harbor of San Julian. The seamen looked at each other, frightened.

But after the *Golden Hind* was free, Fletcher found that he too was in deep water. Francis Drake—adventurer, master mariner, and reader of the psalms—excommunicated his chaplain.

The following selection, from some anonymous notes about the voyage (Vaux, W. S. W., ed., *The World Encompassed*), narrates this episode.

Drake excommunicated Fletcher shortly after that they were come off the rock in this manner. He caused him [Fletcher] to be made fast by one of the legs . . . and a staple knocked fast into the hatches in the forecastle of his ship. He called all the company together, and then put a . . . lock about one of his [Fletcher's] legs.

And Drake sitting cross-legged on a chest, and a

pair of pantofles [slippers] in his hand . . . said, "Frances Fletcher, I do here excommunicate thee out of the Church of God . . . and I denounce thee to the devil and all his angels."

And then he [Drake] charged him [Fletcher] upon pain of death not once to come before the mast, for if he did, he swore he should be hanged. And Drake caused a poesy to be written and bound about Fletcher's arm, with charge that if he took it off he should then be hanged. The poesy was, "Frances Fletcher, the falsest knave that liveth."

HOMEWARD BOUND

Just before he struck the reef, Drake thought he had navigated below Celebes—but he had not. The coast turned south, and once more the northeast monsoon pressed the *Golden Hind* toward a lee shore. Drake beat against the monsoon on short tacks; almost imperceptibly, he drifted south.

On January 20, 1580, Drake reached the Banda Sea. Then he passed through a gap in the barrier of islands, at Timor, and at last, on February 16, 1580, he found himself in the Indian Ocean.

He would not become overconfident again! In business-like fashion, he sailed west along the mountainous southern coast of Java to Tjilatjap, where he anchored March 11. In this port he cleaned his ship for the last time, and he traded with five rajahs clad in turbans and gay-colored silk garments.

One day while he was there, a Portuguese, rowed by natives, came out to the ship.

"*Bien venido,*" said Drake, inviting him aboard. "Welcome."

He showed his visitor the ship—but not the cargo. Casually, Drake led the alert European past big bronze cannon and stacks of pikes, arquebuses, fire bombs, pistols, and corselets.

The Portuguese talked politely, but his face fell. Drake smiled to himself, for he had guessed the truth. The visitor was a spy from a nearby Portuguese trading post whose officials wished to seize the English intruders. They changed their minds after hearing about Drake's arms.

On March 26, 1580, having taken on large supplies of food and water, Drake set sail. Slowly, the high land of Java sank astern. The sandalwood, spices, birds of paradise, and brown people of the East disappeared below the horizon, and the *Golden Hind* sped southwest toward the Cape of Good Hope.

The following selection, from *The World Encompassed* (1628), describes the last leg of the around-the-world voyage of Sir Francis Drake.

We passed on to the Westward . . . till the 9 of *March*, when in the morning we espied land, some part thereof very high. . . . Here we anchored that night, and the next day weighed again, and bearing . . . nearer shore, we came to anchor the second time.

The eleventh of *March* we first took in water, and after sent our boat again to shore, where we had traffic [trade] with the people. . . . The same day, we brought our ship more near the town, and . . . settled ourselves there that night.

The next day our General sent his man ashore to present the king with certain cloth, both linen and woolen, besides some silks, which he [the king] gladly and thankfully received, and returned rice, coconuts, hens, and other victuals. . . . This Island we found to be the Island *Java*, the middle whereof stands in 7 deg. and 30 min. beyond the equator [south latitude].

The 13 of *March*, our General himself, with many of his gentlemen and others, went to shore, and presented the king . . . with his music, and shewed him the . . . use of arms, by training his men with their pikes and other weapons . . . before him. . . .

In this Island there is one chief, but many undergovernors, or petty kings, whom they call *rajahs*. . . . The 14 day we received victuals from two of them, and the day after that . . . three of these kings . . . came aboard to see our General, and to view our ship and warlike munition. . . .

[The next day] the chief king . . . also came aboard. Few were the days that one or more of these kings did [not] . . . visit us . . . whom our General always entertained with the best cheer that we could make. . . .

Though our often giving entertainment in this manner . . . made us spend the more days about them, yet there we found all such convenient helps, that . . . we at last [were ready to leave]. . . . The matter of greatest importance which we did (besides victualling) was the new trimming and washing of our ship, which . . . was . . . overgrown with a kind of shell-fish sticking fast unto her. . . .

We . . . departed . . . the 26 of *March*, and set our course West South West, directly towards the Cape of Good Hope, and continued without touch of aught but air and water, till the 21 of *May*, when we espied land, to wit, a part of the main[land] of *Africa*, in some places very high, under the [south] latitude of 31 deg. and half.

We coasted along till *June* 15, on which day, having very fair weather, and the wind at Southeast, we passed the Cape itself so near in sight, that we had been able with our pieces [cannon] to have shot to land.

July 15, we fell with the land again about *Rio de Sesto*, where we saw many negroes in their boats a fishing, whereof 2 came very near us, but we cared not to stay. . . .

The 22 of the same month, we came to *Sierra Leone*, and spent two days for watering in the mouth of *Tagoine*, and then put to sea again; here also we had oysters, and plenty of lemons. . . .

The 22 day [of August] we were in the height [latitude] of the Canaries. And the 26 of *Sept.* (which was Monday in the . . . reckoning of those that had stayed at home . . . but in our computation was . . . Sunday) we . . . arrived at Plymouth . . . after we had spent 2 years 10 months and some few odd days besides, in . . . going through with so many strange adventures, in escaping out of so many dangers, and overcoming so many difficulties in this our encompassing . . . and passing round about the world, which we have related.

Epilogue

As Drake guided the *Golden Hind* into Plymouth Sound, past the island where he had spent a miserable winter as a refugee, he passed a fishing boat.

"Is the Queen alive?" Drake cried out.

"Yes!" shouted the astonished fishermen.

Drake breathed a sigh of relief. Were Elizabeth no longer upon the throne, he might have had to give up the gold and silver.

The fishermen told Drake that the plague was raging through Plymouth, so he anchored in the harbor and did not land. Then a boat rowed out to the salt-encrusted *Golden Hind*, and Drake's wife and the mayor of Plymouth came aboard.

Later, Drake learned that Mendoza, the new Spanish ambassador, was raging through London, demanding that the treasure in the *Golden Hind* be restored to Spain. Otherwise, Mendoza threatened, Philip II would seize all English shipping in Spanish ports.

Some Privy Council members supported Mendoza; others leaned toward Drake. The Queen, as usual, wavered. First, she ordered the treasure brought up to the Tower and registered—in case she should be forced to return it. Then she decided it should be tucked away in obscure Saltash Castle in Devon—the less said about it the better.

Elizabeth didn't want trouble with Philip II. But neither did she wish to give the gold and silver back—especially after Drake showed her a sample in London and she discovered how much there was.

To Mendoza, Elizabeth pretended that there was *not* very much. And what about the Spanish soldiers who had recently landed in Ireland to stir up rebellion against her? (They were Italians, not Spaniards, and Elizabeth knew it —but she didn't let Mendoza know she knew.)

By April 4, 1581, Elizabeth had made up her mind. She was still fencing with Mendoza, but Philip II had taken no action. She had ordered the *Golden Hind* taken to Greenwich, on the Thames, where she planned to preserve it as a national monument. Now she journeyed there in state and knighted Drake on his own ship after an elaborate banquet.

"I have a gilded sword with which to strike off your head," the Queen said merrily—and then made her captain "Sir Francis."

The English people had already given their accolade to their pirate-hero, the commoner and true-blue Protestant who had defied Pope and Emperor and won a fortune to boot. Drake was the Charles Lindbergh or John Glenn of the day. Songs about him blared in all the taverns. Hawkers sold pictures and books about him, and crowds followed him through the streets.

"The master thief of the unknown world," one disgruntled critic called him, but the critic was in the minority. Most Englishmen thought that snatching several million dollars in gold, silver, and jewels from the King of Spain was a glorious feat.

Of this sum, Drake had been given fifty thousand dollars for himself and another fifty thousand dollars to distribute among his crew, and he undoubtedly received more later. The best Mendoza could get was permission for Spanish merchants to sue for damages in English courts, where most of the suits were dismissed on technicalities. Philip II did not go to war—yet.

He did not think he would have to. By 1581, things were going so well for Spain that it appeared Philip would soon outflank England and reduce it to impotence without formal hostilities. For in spite of Drake's robberies and English military aid to the Netherlands rebels, the Duke of Parma was rapidly subduing the Low Countries for Spain. In 1578, Philip had reached an agreement with the dominant French faction and no longer had to fear that nation. In 1580, he inherited the throne of Portugal and the far-flung Portuguese maritime empire.

Even cautious Elizabeth was alarmed. Could she afford to have an all-powerful Spain, unchecked by France, astride the Netherlands only a few miles from the English coast? As Philip's infantry began to seem unconquerable, there remained just one force Elizabeth could turn to: the greed and boldness of English mariners, which had been demonstrated in Drake's famous voyage around the world.

In 1585, relations between England and Spain reached a crisis. Just when Elizabeth was desperately trying to negotiate a Netherlands settlement and while an unusually large number of English ships were in Spanish ports, Philip ordered every English vessel seized. Instantly, English merchants trading in Spain, who had hitherto been "doves," turned against Spain. Elizabeth increased military aid to the Netherlands—and sent for Drake.

Sir Francis had occupied the intervening years buying a splendid estate (Buckland Abbey), serving as mayor of Plymouth, getting elected to Parliament, and planning a second expedition to the Spice Islands. At Elizabeth's request, he now changed his plans and began preparing for a major raid on the West Indies.

In September, 1585, Drake sailed with 22 ships and 2,300 men to ravage the Caribbean. His old employer John Hawkins, now treasurer of the navy, had greatly strengthened Elizabeth's navy by building a number of the new "race ships" and remodeling older vessels. With the aid of some of these ships, Drake planned to attack Panama and to seize and hold at least one town as a base—perhaps Havana.

The fleet sacked Santo Domingo and Cartagena, but because of a fever that decimated the crews, it did not get to Panama. On the way back, Drake also destroyed Spanish St. Augustine, Florida. He stopped at the island of Roanoke, in North Carolina, the first English colony in the New World, where he found the settlers so discouraged that he carried them all back to England. This colony had been founded earlier in the year by Richard Grenville, primarily as a base from which to attack the Spanish Indies.

Drake brought back about $150,000 in booty, chiefly ransom from towns he had captured. It was less than he had hoped for, but the raid had an impact out of proportion to the treasure seized and the damage done.

It hurt Spanish prestige throughout the world, thoroughly frightened the Spanish colonists, and made a reluctant Philip decide to take a gamble on invading England. After Elizabeth and Drake's "Atlantic offensive" of 1585, Philip saw that English sea power threatened to make pacification of the Netherlands impossible. The Netherlands were essential to Spanish prosperity—hence Philip's fateful choice.

A great Spanish admiral, the Marquis of Santa Cruz, began to assemble the Armada in 1586, and the initiative in the undeclared war now passed to Spain. Elizabeth had to strain every nerve to protect the home country; privateers and naval vessels could no longer be unleashed to prey on Spain's New World possessions.

In April, 1587, Drake led a fleet of twenty-three ships to the coast of Spain. He was ordered not only to take prizes at sea but also to disrupt Santa Cruz's preparations by entering ports and destroying the shipping there. It was not quite a case of an ex-robber joining the police force, but at least Drake the pirate had to change into Drake the admiral and naval strategist.

"The wind commands me away," wrote Drake. "Our ship is under sail. God grant we may so live in His fear as the enemy may have cause to say that God doth fight for her Majesty."

As soon as he reached Spain, Drake learned that the harbor at Cadiz was full of ships and was almost unprotected. With a pirate's opportunism, he passed up the consultation with captains called for by naval custom and drove his fleet straight into the port. Here he destroyed or captured between twenty-four and thirty vessels, including a splendid galleon belonging to Santa Cruz himself. He called it "singeing the king of Spain's beard."

Then he took Sagres Castle near Cape St. Vincent, which commands the route from the Mediterranean coast of Spain to the Atlantic. From this lair he seized a number of coasters carrying staves which were to be made into water casks for the Armada. This incident led to there being some very thirsty Spanish sailors later, for Santa Cruz had to replace the lost staves with green ones that caused the casks to leak. Drake also sailed to Lisbon, where Santa Cruz was readying his chief ships, and challenged the Spanish admiral to come out and fight—but Santa Cruz could not, or would not.

On May 17, 1587, Drake wrote a member of Elizabeth's Council that he planned to stay at Cape St. Vincent for weeks, harrying the Spaniards—yet five days later he rushed off under full sail for the Azores. He had heard of a rich East Indian carrack approaching those islands, and Drake the pirate promptly overcame Drake the admiral.

His luck was incredible. He not only captured the carrack with goods valued at twice the amount brought back from the West Indies raid, but by pure accident he drew Santa Cruz after him. Santa Cruz feared for the treasure fleet from Panama, which was also nearing the Azores.

The Spanish admiral never saw Drake, and by the time he returned to Lisbon his ships and men were in such bad repair that the invasion had to be postponed until the spring of 1588. Before that, Santa Cruz died, and a conscientious but less capable man, the Duke of Medina-Sidonia, filled his position.

The showdown came in 1588. By April, Medina-Sidonia had assembled 130 ships—a "power" fleet with heavy, short-range cannon. The towering galleons were crowded with soldiers who would board the enemy, two soldiers for every sailor. The English force, on the other hand, was a "navy man's navy," which included many low, fast race ships, was superior in long-range artillery, and carried more seamen than soldiers. It was commanded by the tactful, intelligent Lord High Admiral Charles Howard, a strong supporter of Hawkins; Drake was the vice-admiral.

The Spaniards would sail in a bulky "eagle" formation and try to crash through the Channel to Dunkirk, where the Duke of Parma's invasion army from Flanders would come on board. The English would use a line-ahead formation, stringing their vessels out like Indian scouts and striking from a distance to disrupt the Spaniards.

In April, 1588, Drake urged the Queen to allow him and Howard to attack the Armada off the coast of Spain. Permission was granted, and in June the English force sailed almost to within sight of Coruña, where the Armada was gathered. Suddenly, the wind shifted and blew the English back to the Channel.

The fleet then anchored in the harbor at Plymouth. One day not long afterward—July 19, 1588—Drake, Howard, and other officers were bowling on a green when a distraught captain rushed up and cried, "Gentlemen, the Armada is entering the Channel!"

The officers were stunned, but Francis Drake picked up his bowling ball. "Finish the game," he said calmly. "We have time enough to finish the game and beat the Spaniards afterwards."

Then, while beacon fires flared from headland to headland, the English went into action. They were in a dangerous position. The same wind which had carried their fleet away from Spain had brought the Armada sooner than expected to the western end of the Channel, which it filled like a stopper in a bottle. The English might be trapped in Plymouth.

All that night, they worked their fifty-four ships against the wind, out into the Channel. The next afternoon, through mist and drizzle, they caught their first sight of the Armada. It was in close formation, moving very slowly like a caravan of heavily-burdened elephants. This second night, July 20, by skillful tacking, Drake and Howard led their ships around and behind the Spaniards.

Now Drake had the windward position. He was west of the Armada and could follow it through the Channel, bombarding it and shepherding it away from the coast.

For a week the English pursued the Armada, the English force increasing to nearly one hundred ships. On July 21, however, Drake the pirate may have returned briefly. That evening the entire English fleet was following Drake's poop lantern as he led them after the Armada when, suddenly, the lantern went out. Some ships stopped; others, looking for the lantern, increased speed and nearly tailgated the Armada. The next day Drake reappeared, having captured a rich Spanish prize during the night!

When the two fleets came opposite Calais, near where Medina-Sidonia hoped to encounter Parma's army, Medina-Sidonia made his second mistake. (His first had been letting the English slip past him to the windward position.) Instead of seizing an English port as a base, he now anchored his fleet off Calais—July 27, 1588—so that the galleons became sitting targets.

The next night, Drake and Howard launched eight fire ships against the immobilized Armada. At the sight of these bonfires floating toward them, the Spaniards panicked and cut their cables. As their ships swung around in hopeless confusion, Drake and Howard attacked.

For twelve hours, guns blazed, smoke billowed from ports and fire ships, shouts and the roar of cannon filled the air. Two Spanish galleons were reduced to wreckage. One Spanish great ship sank. Drake charged Medina-Sidonia's flagship, the *San Martin*, and several other galleons were isolated from the main body of the Armada and surrounded. Suddenly, driving rain descended like a curtain between the vessels, and the battle of Gravelines ended.

The Spaniards had already begun to re-form. When the squall passed, they regained their impregnable "eagle" formation. But they had been carried past Calais, past Dunkirk, past all hope of embarking Parma's army and conquering England. Short of anchors because of the

An old German print showing the Spanish Armada in the English Channel the night before hostilities began.

cables they had cut, facing the indomitable fleet of Drake and Howard as well as the boisterous south wind, they could not get back to the Channel or Flanders.

Their fleet was driven into the North Sea. Drake and Howard followed for several days, then turned away to the east coast of England. But the storm-battered Armada, now also riddled with disease, sailed on—over the top of Scotland and around Ireland on a terrible homeward voyage. Fifty of its original 130 ships and thousands of its men were lost before it reached Spain.

After the defeat of the Spanish Armada, during which English privateers, naval men, and ex-pirates had risen above themselves to organize victory against an "invincible" foe, many captains relapsed into the free-lancing of earlier days. The war with Spain dragged inconclusively on. But the English overseas expansion also continued.

England, now safe from invasion, would be able to acquire an empire. She had won at least the freedom of the seas—equality with Spain and an option on a continent with a great future, North America. She could now plant and maintain colonies there—as she could not before—because her navy had wrested control of the North Atlantic from Philip II.

In all this, Drake had been a brilliant leader and a symbol. After the circumnavigation, his fame extended throughout Europe. The greatest of Spanish playwrights, Lope de Vega, wrote an epic *La Dragontea* ("The Epic of the Dragon") about his feats. Many princes of Italy and Germany desired his picture. Foreign aristocrats as well as English commoners eagerly sought scraps of information about him.

It did not matter that Drake's final voyages, attacks on Lisbon (1589) and Panama (1595–96), ended in failure. It seemed almost more pertinent that among Drake's last words were ones characteristic of the great corsair who had captured the world's imagination. Dying of dysentery, January 28, 1596, off the Panama which had inflamed his desires ever since his first raid on the Spanish Main, Drake cried, "We must have gold!"

Drake had, indeed, found gold for England—not only the glittering bullion of Peru, but also the more lasting treasure of trade routes to the East and sea-lanes across the Atlantic to the New World. In the next century, as Spain began its slow decline, Englishmen would market goods in Bombay and build homes in Virginia and Massachusetts. Merchants, courtiers, and above all the mariners who loved Drake—the new social class among the poor—would follow his noble steps overseas and in so doing relieve unemployment and avoid strife in the nation.

They, and their descendants, would be inspired by voyages which were the supreme expression of Elizabethan ambition, individualism, and daring—the voyages of the sea-dragon, Sir Francis Drake.

Bibliography

SOURCES FOR DRAKE'S VOYAGES

Arber, Edward, *An English Garner*, Vol. IV, *Voyages and Travels*, C. R. Beazley, ed. New York, E. P. Dutton and Company, 1877–1890.

Hakluyt, Richard, *The Principal Navigations . . . of the English Nation*, 12 vols. London, J. M. Dent & Sons, Ltd., 1926–1931.

Nuttall, Zelia, ed. and tr. *New Light on Drake*, No. xxiv, Second Series, The Hakluyt Society, London. Published by Cambridge University Press (London), 1914.

Vaux, W. S. W., ed., *The World Encompassed by Sir Francis Drake* . . . London, The Hakluyt Society, 1854.

Wright, Irene A., ed. and tr., *Documents Concerning English Voyages to the Spanish Main, 1569–1580*. London, The Hakluyt Society, 1933.

BOOKS ABOUT DRAKE

Andrews, Kenneth R., *Drake's Voyages*. New York, Charles Scribner's Sons, 1968.

Benson, E. F., *Sir Francis Drake*. New York, Harper & Brothers, Publishers, 1927.

Corbett, Julian S., *Drake and the Tudor Navy*, 2nd ed., 2 vols. New York, Burt Franklin, 1965. (First printed 1899).

Mason, A. E., *The Life of Francis Drake*. New York, Doubleday, Doran & Company, Inc., 1942.

Wagner, Henry R., *Sir Francis Drake's Voyage Around the World*. San Francisco, John Howell, 1926.

Williamson, James A., *The Age of Drake*. London, A. & C. Black Ltd., 1938.

Williamson, James A., *Sir Francis Drake*. Hamden, Connecticut, Archon Books, 1966. (First printed 1951).

BOOKS ABOUT DRAKE'S TIMES

Bourne, Edward Gaylord, *Spain in America, 1450–1580*. New York, Harper & Brothers, Publishers, 1904.

Elliott, J. H., *Imperial Spain, 1469–1716*. New York, St. Martin's Press, 1964.

Elton, G. R., *England Under the Tudors*. London, Methuen & Co. Ltd., 1955.

Gosse, Philip, *Sir John Hawkins*. London, John Lane, The Bodley Head, 1930.

Kingsley, Charles, *Westward Ho!* New York, Macmillan and Co., 1886. (First printed 1855).

Penrose, Boies, *Travel and Discovery in the Renaissance, 1420–1620*. Cambridge, Harvard University Press, 1955.

Sauer, Carl O., *The Early Spanish Main*. Berkeley, University of California Press, 1966.

Smith, Lacey Baldwin, *The Elizabethan World*. New York, American Heritage Publishing Co., Inc., 1967.

Taylor, E. G. R., *Tudor Geography, 1485–1583*. London, Methuen & Co. Ltd., 1930.

Williamson, James A., *Hawkins of Plymouth*. London, A. & C. Black Ltd., 1949.

Cast of Characters

San Juan de Anton. Unsuspecting captain of a Spanish treasure ship, the *Cacafuego*, seized by Drake off the coast of Peru.

Sultan Baber. Cunning ruler of the island of Ternate in Indonesia. Entertains Drake lavishly and agrees to trade spices with the English—but also considers killing Drake by treachery.

Lord Burleigh. Lord Treasurer of England, Elizabeth's chief adviser. Leader of the party seeking peace with Spain. Stout and stout-hearted; a sage, experienced diplomat.

Miguel de Castellanos. Treasurer of Rio de la Hacha on the north coast of South America. Bravely resists Drake, but is betrayed by a slave.

John Cooke. Member of Drake's expedition to the Pacific; dislikes Drake. Writes a vivid "Narrative" of the voyage, giving the only detailed account of the trial and execution of Thomas Doughty.

John Dee. White-bearded geographer and magician interested in voyages to the New World. Believes in the existence of Terra Australis, a vast, treasure-filled southern continent.

Thomas Doughty. Gentleman-adventurer who helps Drake plan his voyage to the Pacific. Imaginative, well-educated, conceited. Accused of treason and executed by Drake in South America.

Edmund Drake. Francis Drake's father—a tenant farmer, later a navy chaplain. An ardent Protestant who reads the Bible aloud to his twelve sons and gives Francis his robust faith.

Francis Drake. Master mariner who leads the English advance into the New World. Called "the great pirate" by the Spaniards. Victor over the Spanish Armada and circumnavigator of the globe.

John Drake. (*1*) Francis Drake's capable but rash brother who commands a ship in the 1572 expedition to Panama. Loses his life in an attack on a heavily armed Spanish vessel.

John Drake. (*2*) Francis Drake's young cousin who serves as his page on the voyage around the world. Later falls into the hands of the Inquisition and gives two depositions describing the voyage.

Joseph Drake. Another of Francis Drake's brothers, a seaman on the 1572 expedition. Dies of a fever on the Isthmus of Panama in 1573.

Thomas Drake. Francis Drake's youngest brother, who lives to be his heir. A member of the expedition around the world; accused of theft by Thomas Doughty, but defended by Francis Drake.

Queen Elizabeth I. "Gloriana"–slender, with red-gold hair and dark eyes. Cautious, changeable, plain-spoken, and unexpectedly daring. Seeks both peace and power for England.

Don Martin Enriquez. Viceroy of Mexico. Energetic and proud, a determined opponent of Drake and other English intruders in Spain's New World empire.

Francis Fletcher. Moralizing chaplain of the *Golden Hind.* Writes a rather wordy but detailed "Narrative" of the voyage around the world, used by the compiler of *The World Encompassed.*

Sir Christopher Hatton. Handsome Captain of the Guard and a favorite of Elizabeth's, noted for his elegant dancing. Supports Drake's voyage to the Pacific. Drake's flagship is renamed the *Golden Hind* in honor of Hatton.

John Hawkins. Daring and persuasive merchant-mariner of Plymouth. Begins English voyages to the Caribbean, forcing his goods on the Spaniards there.

Lord Charles Howard. Admiral of the English fleet that fought the Spanish Armada. Tactful, generous, but something of a figurehead—leans heavily on Drake for counsel.

Earl of Leicester. Elizabeth's first favorite, a swarthy, dashing adventurer nicknamed "the Gypsy." Leader of the "war party" at court and backer of Drake's raids against the Spanish empire.

Duke of Medina-Sidonia. Commander of the Spanish Armada. A good Christian who conscientiously does his duty but prefers his orange groves to the sea.

John Oxenham. A lesser Drake. Fanatically anti-Spanish and anti-Catholic, daring to the point of recklessness. Tries to seize the Isthmus of Panama with only a handful of followers, but is captured and later hanged in Lima.

Philip II. King of Spain—Elizabeth's former brother-in-law and her great rival. Austere, humorless, an indefatigable worker. Views Spain as the champion of the Catholic Counter-Reformation.

Pedro Sarmiento de Gamboa. Outstanding Spanish navigator and humanist who had trouble with the Inquisition. Later took part in the discovery of the Solomon Islands. Participates in the pursuit of Drake and writes a lively "Narrative" about the chase.

Nuño da Silva. A short, bearded Portuguese pilot captured by Drake off the Cape Verde Islands. A skillful navigator who helps Drake sail the *Golden Hind* into the Pacific and is indignant when left behind in Mexico.

Don Luis de Toledo. Viceroy of Peru, a fiery veteran who quickly organizes the pursuit of Drake after the raid on Callao, but forgets to warn the treasure ship, the *Cacafuego*, of Drake's presence.

Captain Tetû. Sad French Huguenot captain who is given supplies by Drake and, in turn, helps Drake capture a Spanish treasure-train near Nombre de Dios. Is fatally wounded in the fight.

Mary Tudor. Over-zealous Queen of England who hopes to persuade her countrymen to return to Catholicism but is remembered as "Bloody Mary" because of her later persecution of Protestants.

John Winter. Enigmatic captain of the *Elizabeth*, who sails back to England from the Strait of Magellan, leaving Drake alone in the Pacific. In a court deposition, he blames Drake for making piratical attacks on the Portuguese.

Don Francisco de Zarate. Spanish nobleman captured by Drake in the Pacific, off Mexico. A keen observer of Drake's ship and men.

A Timetable of Events

October 2, 1567 Drake sails, under John Hawkins, on a "troublesome voyage" to the Spanish Main.

Spring, 1568 Bombards Rio de la Hacha, a Spanish town on the northern coast of South America.

September 23, 1568 Is attacked along with the rest of Hawkins' force by a Spanish fleet at San Juan de Ulua, Mexico, but escapes.

1570, 1571 Makes reconnoitering voyages to the Isthmus of Panama, perhaps entering Nombre de Dios as a spy, and plans to seize Spanish treasure.

May 24, 1572 Sails from Plymouth, England, with two ships and seventy-three men, on a raid against Panama.

July 29, 1572 Seizes and holds Nombre de Dios for a few hours but fails to obtain treasure; he is wounded and retreats.

February 11, 1573 Sees the Pacific Ocean for the first time, from a "great high Tree" on the Isthmus, and vows to sail on it.

April 1, 1573 Waylays a Spanish mule-train (*recua*) outside Nombre de Dios, and takes booty worth two hundred thousand dollars.

August 9, 1573 Arrives in Plymouth with his treasure but goes into hiding because relations between Spain and England have improved.

April, 1575 Reappears in Ireland, serving in an English expedition against the Irish; meets gentleman-adventurer Thomas Doughty; they plan a voyage to the Pacific.

Spring (?), 1577 Has an interview with Queen Elizabeth, who says she "would gladly be revenged on the King of Spain" through Drake's forthcoming voyage.

November 15, 1577 Departs from Plymouth with five ships and 140–164 men on a mysterious expedition headed toward the Strait of Magellan.

January 30, 1578 Captures a Portuguese caravel off the Cape Verde Islands; retains its veteran pilot, Nuño da Silva.

July 2, 1578 Executes gentleman-adventurer Thomas Doughty for treason at Port San Julian, Argentina.

August 20–September 6, 1578 Navigates the Strait of Magellan with three ships: the *Golden Hind*, the *Elizabeth*, and the *Marigold*.

September 7–October 19, 1578 Is driven far south in the Pacific, near Cape Horn, by a storm; the *Elizabeth* and the *Marigold* disappear.

November 26, 1578 Is ambushed and wounded by Arauncanian Indians on the island of Mocha, off Chile.

December 5, 1578 Enters the harbor of Valparaiso, Chile, revealing his presence in the Pacific to the Spaniards.

February 13–14, 1579 Attacks shipping in Callao, the port of Lima, Peru; perhaps tries to rescue John Oxenham, a prisoner in Lima, but fails.

March 1, 1579 Captures a treasure ship, *Nuestra Señora de la Concepción* (the *Cacafuego*), off Ecuador, containing gold and silver worth several million dollars.

March 6–16, 1579 Escapes pursuing Spanish ships by sailing north to Nicaragua.

April 13–16, 1579 Raids Guatulco in southern Mexico, abandons the Portuguese pilot da Silva there, and plans the return voyage to England.

June 17–July 23, 1579 Establishes a base on the coast of California, just north of San Francisco, and overhauls the *Golden Hind*.

June 24, 1579 Is offered a "kingdom" by the California Indians; he accepts in Queen Elizabeth's name and calls the land "Nova Albion."

July 23–September 30, 1579 Crosses the Pacific Ocean, without incident, to the Pelew Islands near the Philippines.

November 3–9, 1579 Makes a trade treaty with Sultan Baber of Ternate, in Indonesia, and takes on a cargo of spices.

January 9–10, 1580 The *Golden Hind* runs aground on a reef in the Molucca Sea but is freed by a shift of wind.

March 11–26, 1580 Drake trades with rajahs at Tjilatjap, on the southern coast of Java, and prepares for the homeward voyage.

June 15, 1580 Rounds the Cape of Good Hope.

September 26, 1580 Anchors in the harbor at Plymouth, having sailed around the world in a little less than three years.

April 4, 1581 Is knighted by Queen Elizabeth at Greenwich, on the Thames.

September, 1585–July, 1586 Commands 22 ships and 2,300 men in a major raid on the Caribbean.

April–June, 1587 Harasses the coast of Spain, destroying much shipping in the harbor at Cadiz and seizing barrel staves intended for the Spanish Armada.

July 19–August 2, 1588 Serves as vice-admiral in Lord Howard's fleet, which successfully defends England against the Armada.

January 28, 1596 Dies of dysentery, off Panama, on a final raid against the Spanish Empire.

PICTURE CREDITS

Title page	Rare Book Division The New York Public Library Astor, Lenox and Tilden Foundations
page xxiii	Prints Division The New York Public Library Astor, Lenox and Tilden Foundations
pages xxxi, 9, 22, 42–43	Rare Book Division The New York Public Library Astor, Lenox and Tilden Foundations
page 45	National Portrait Gallery, London
pages 61, 64, 83, 97, 111, 112, 133, 169	Rare Book Division The New York Public Library Astor, Lenox and Tilden Foundations
page 176	By permission of the Bancroft Library, University of California, Berkeley
pages 193, 215	Rare Book Division The New York Public Library Astor, Lenox and Tilden Foundations

Index

Alexander VI, Pope, xxvi
Alexandria, 48
Almiranta, 124–25, 141
America, discovery of
  English-Spanish alliance, influence on, xxvi
Angel, xxxii, xxxviii
Angel, Miguel, 121, 123, 149
Anian, Strait of (*see* Northwest Passage)
Antony, Nicholas, 75
Araucanian Indians, 95–100, 229
Arequipa, 114, 116
Arica, 110, 113, 114–16
"Atlantic offensive," 210
Auto-de-fé, xxvii
Azores, 212

Baber, Sultan, 187–88, 221, 238
Bachian, 187
Balboa, Vasco Núñez, 17
Banda Sea, 201
Baptista, Alonso Rodriquez, 122
Benedict, 46, 47, 55
Borburata, xxxiii
Brava, 54–55
Brewer, John, 58 59, 88, 89
Bright, Edward, 58, 68, 88
Buckland Abbey, 209
Burleigh, Lord, xxiv–xxv, 40, 66–68, 221

Cabo Virgen Maria (*see* Cape Virgins)
Cabot, John, xxvi
Cabrillo, 173
Cacafuego, 121, 126–34, 221, 229
Cacaplata (*see Cacafuego*)
Calais, 214
California
  Drake in, 168–73, 230
  Indians of, 168–73, 230
Callao, 117–18, 123, 126
Calvin, John, xxi
Caño, 132, 144
Cape Deseado, 85–87, 89
Cape of Good Hope, 202–5, 230
Cape St. Antonio, 30–31
Cape St. Vincent, 211–12
Cape Verde Islands, 52–55, 228
Cape Virgins, 78–79
Capitana, 124, 141
Caracoa, 190
Caribbean, xiii, xxvi, 36, 230
Cartagena, xxxiv, 210
Castile, 17
Catholicism, xx, xxvi
Celebes Island, 192–96
Celebes Sea, 184
Chancellor, Richard, xxviii
Charles V, Emperor, xxi
Chester, Captain John, 47, 65
Chile, 39, 94–95, 102–9
Cimaroons, 7, 15, 18–20, 29–31

Colchero (Spanish pilot), 144–47, 151, 153–54
Colonialism, xxx
Columbus, Christopher, 91
Cooke, John, 57–60, 66–71, 221
Corço, Felipe, 115
Cortes, Hernando, xxii, 14
Crab Island, 192, 194–96
Crowndale, xiv–xv
Cuttill, 63
Cuzco, 110, 113

Da Silva, Nuño, 51, 53–55, 60–63, 114, 154–55, 179, 224–25, 228, 229
Davis, John, xxix
De Abreu, Antonio, 17
De Anton, Captain San Juan, 117, 126–32, 134, 137, 221
De Arana, Pedro, 125
De Castellanos, Miguel, xxxiii–xxxiv, 221
De Dueñas, Francisco, 184–87
Dee, John, xxviii, 35, 221–22
De Eraso, General Don Cristobal, 158
De Frias Trejo, Diego, 123, 139, 148
De Gamboa, Pedro Sarmiento, 104, 121–26, 127–32, 136, 139, 140, 148, 224
De Neyra, Mendaña, 104
De Ortega, Pedro, 139
De Toledo, Don Luis, 120–23, 148, 225
De Vargas, Gaspar, 155–58
De Vega, Lope, 216
De Velasco, Don Luis, 162–64
De Zarate, Don Francisco, 146–47, 149–54, 194, 225
Dominica, 4
Doughty, John, 72
Doughty, Thomas, 38–39, 52, 55–58, 221, 222, 228
  defense of, 63, 69–71
  disgrace of, 60
  execution of, 71–72, 228
  Lord Burleigh and, 66–68
  treason of, 58–60, 66–68
  trial of, 65–71
Drake, Edmund (father of Francis), xiv–xv, xvii, xx, 222
Drake, Francis
  ability of, 37
  Acapulco, attack plans for, 149, 154
  as admiral and strategist, 211–16
  apprenticeship of, xxi
  Arequipa, attempt to raid, 116
  in Argentina, 64
  Arica, raid on, 115–16
  Atlantic crossing, 52–63, 65
  "Atlantic offensive" of, 210–11
  in Azores, 212
  Baber, Sultan, and, 187–88, 221, 230
  in Banda Sea, 201
  birth of, xiv
  Cacafuego, capture of, 126–33, 221, 229
  in California, 166–73, 175–77, 180, 229, 230
  at Caño, 144
  Cape of Good Hope, rounding of, 230
  in Cape Verde Islands, 52–55

Drake, Francis (*Cont.*)
Cape Virgins, arrival at, 78–79
Caribbean raid, 230
Catholicism, opposition to, xx, xxvii
in Celebes Sea, 184, 196
in Chile, 94–95, 102–9
church destruction by crew of, 159–60
Cimaroons, aid to, 15
circumnavigation
equipment for, 44–48
final stage of, 201–5
first attempts at, 48
map of, 42–43
Colchero and, 145–47, 153–54
at Crab Island, 192, 194–96
crew of, 46, 72–76, 94–100
Da Silva, Nuño, and, 53–55, 63
death of, 217, 231
De Gamboa, Pedro Sarmiento, and, 139–40
description of, 2, 222
at Devon, xiv–xv
De Zarate, Don Francisco, and, 150–52
disappearance of, 34
Doughty, Thomas, and, 55–58, 65–71, 228
as *el gran corsario*, xiv
Elizabeth I, Queen (England), and, 41, 44
England, reception in, 208–9
financial backing of, 38, 44
first voyage of, xxiv, xxxii–xxxviii, 227
Fletcher, Francis, and, 200–1
fourth voyage of, 216–17

Drake, Francis (*Cont.*)
geography, revision of theories on, 92–94
at Guatulco, 154–58, 165, 229
Hatton, Christopher, and, 78, 79
Hawkins, John, and, xxii–xxiv, xxxiii–xxxiv
history, place in, 217
in Ireland, 34, 228
kidnapping, accused of, 48
knighting of, 208, 230
at Lima, 117
maps of, 86, 105–6, 147
Magellan, Ferdinand, comparison to, 180
Magellan, in Strait of, 78–100
Martin, Gaspar, and, 117
at Mocha Island, 94–100
naval training of, xix, xxi
Nicaragua, departure for, 129
at Nombre de Dios, 7–14, 227, 228
at Nova Albion, 173–76
Oxenham, John, and, 37, 117–20, 130
in Panama, xiii, 15–17, 23–28
Patagonian Indians and, 65
in Pelew Islands, 180–81, 230
Pelican, as admiral of, 47
Peru, arrival at, 109–10
in Plymouth (England), 31, 207–9, 228
Polynesians and, 181–83
Port Pheasant landing, 6–7
reappearance of, 38
religion and, xv, xvii, xx
at Rio de la Hacha, xxxiv, 227
at Rio Francisco, 28
at St. Augustine (Fla.), 210

Drake, Francis (*Cont.*)
at San Juan de Ulua, xxxvii–xxxviii, 227
and San Julian, 65–76
and Santa Cruz, 211
second voyage of, 2–31, 227
slave trade and, xxx, xxxii–xxxiii
in South Sea, 38–40, 85, 147, 230
Spain and, xxix, xxx, xxxvii, 3, 211, 229
Spanish Armada, defeat of, 212–16
Spice Islands and, 187–92
superstition and, xxx, xxxii, 60
at Tarapacá, 107–9
at Ternate, 188–93
Tetû, Captain, and, 23
third voyage, 227
at Tjilatjap, 201–2
treason, accusations by, 75
treasure, Spanish, and, 29, 107–9, 134
Valparaiso, entry of, 229
Velasco's theories on whereabouts of, 162–65
in West Indies, xxxiii, 4–5, 210
Winter, Captain John, and, 134–35
in winter quarters, 65, 72
worship, conduct of, 160–61
youth of, xv–xxii
Drake, John (brother of Francis), 2, 8, 11, 13, 15, 222
Drake, John (cousin of Francis), 77, 118–20, 194, 222
Drake, Joseph (brother of Francis), 2, 15, 222
Drake's Bay (*see* Point Reyes)
Drake's Island, xvi–xvii
Drake's Pool, 34
Drake, Thomas (brother of Francis), 57–58, 223

East India Company, 187
East Indies, xiv
Edward VI, King (England), xvi, xx
El gran corsario, xiv
Elizabeth, 46, 47, 55–56, 65, 75, 76, 85, 86, 89, 90, 93, 94, 108, 134, 228–29
Elizabeth I, Queen (England), xxx, 33, 36–38, 45, 210, 223
Burleigh, Lord, and, xxiv, xxv
Drake, Francis, and
knighting of, 208, 230
meeting with, 40–44
support of, 44, 228
as "Gloriana," xxv
Grenville, Richard, and, 35
Netherlands, aid to, 209
Phillip II, King (Spain), and, 34, 39, 207–8
Ridolfi Plot and, 2–3
Elizabeth Island, 94, 95
England
Catholicism in, xx
exports of, xxv
Far East, route to, and search for, xxix
in New World, 34–35
piracy of, 3
Protestantism in, xvi
Russia, trade with, xxviii

England (*Cont.*)
sea power of, xxvi–xxvii, xix
Spain and
as ally of, xxv–xxvi
relationship of, xxxix, 2–3, 35–36
as rival of, xxviii
trade with, 34
Spanish Armada, defeat of, xix, 216
English-Spanish alliance, xxvi
Enriquez, Don Martin, xxxvi, xxxvii, 149, 155–59, 161–62, 164–65, 223
Erasmus, xxvii
Espírto Santo, 63
Essex, Earl of, 38, 39, 41, 69
Europe, commercial rivalry in, xxvi

Fletcher, Francis, 53, 73, 86, 90–91, 200–1, 223
Fogo, 53
Fogo Island, 54
France, xxv, 209
Francis I, King (France), xxvi
Frobisher, Martin, xxix, 35

Galleasses, xvii
Galleon, xvii–xix
Garret, John, 6
Gilbert, Humphrey, xxix, 35
"Gloriana" (*see* Elizabeth I, Queen)
Golden Hind, xxxix, 78–79, 85–86, 89, 91, 94, 107, 109, 133, 136, 147, 154, 177, 192, 197–200, 208, 223, 225, 228, 229, 230
Gomez, Francisco, 156
Gonçalo, Don, 151
Great ship, xvii–xviii
Grenville, Richard, 35–37, 40, 210
Griego, Juan, 120, 125
Guadaloupe, 4
Guatulco, 154–58, 229
Guayaquil, 127

Hakluyt, Richard, xxix
Hatton, Christopher, 38–39, 78, 79, 223
Hawkins, John, xxii, xxiii, 161, 210, 223, 227
as aggressor against Spain, xxix, xxxii
as commander, xxxii
Catholic pose of, xxxiii
Enriquez, Don Martin, and, xxxvi
at Rio de la Hacha, xxxiv
at San Juan de Ulua, xxxv, xxxvi
slave trade and, xxxii–xxxiii
Spaniards, first battle with, xxxvii
Spaniards, trade with, xxxiv
third voyage of, xxx, xxxii–xxxviii
in West Indies, xxxiii
Hawkins, Richard, 91
Hawkins, William, xxvi, 35
"Hawks" of Queen Elizabeth, 35
Henderson Island, 91

Henry VIII, King (England), xvi, xix
Hood, Thomas, 75
Horn Islands, 91
Howard, Lord Charles, 212–14, 216, 223–24, 231
Hudson Bay, 35
Hudson Strait, 35

Indians
  Araucanian, 95–100, 229
  Brazilian, 62–63
  Californian, 168–73, 230
  Fuegian, 82
  Patagonian, 64–65
Island of Thieves, 183
Islas de los Ladrones (*see* Island of Thieves)

Jesus of Lubeck, xxx, xxxii, xxxiv, xxxv, xxxvii, xxxviii
Judith, xxx, xxxii, xxxvii, xxxviii

La Dragontea, 216
La Herradura, 106–8, 116
Latimer, Hugh, xx
Leicester, Earl of, 224
Lima, 113–14, 117
Llamas, 110, 113
London, 44
Lovel, Captain John, xxxiii
Low Countries, 209
Luther, Martin, xxvii
Lutheranism, Catholicism and, xxvii

Madrid, Juan, 155, 157
Magellan, Ferdinand, 17, 65, 79, 180
Magellan, Strait of, 34–39, 49
  Drake's voyage through, 78–100, 228
  islands in, description of, 80–82
  map of, 83
  Peruvian guard at, 121
  Spanish expeditions to, 80
Makian, 187
Manta, 138
Mar del Sur (*see* South Sea)
Margarita, island of, xxxiii
Marigold, 46, 47, 65, 75, 76, 85, 88, 90, 94, 108, 134, 228–29
Markham, William, 75
Martin, Gaspar, 117–18
Mary, 55, 58
Medina-Sidonia, Duke of, 212, 214, 224
Mendoza, Spanish ambassador, 207–9
Menendez, Pedro, xxx
Merchant Adventurers company, xxv
Mindanao, 184
Minion, xxxii, xxxvii, xxxviii
Minivy, Richard, 106–8, 116
Mocha Island, 94–100, 229
Moluccas (*see* Spice Islands)
Molucca Sea, 230

Montezuma, xxii, 14
Moone, Captain Thomas, 47
More, Thomas, xvi, xxvi, xxvii
Mortier, 187
Muscovy Company, xxviii

Naval battles, great ships in, xviii
Netherlands, xxviii, 39, 209
New economics, xv–xvi
Newman, Mary, 3
New Spain, 132, 145, 159
Nombre de Dios, xiii, 5, 7–14, 23–24, 127, 227, 228
Northwest Passage, 35, 39, 165–66
Nova Albion, 173–75
Nuestra Señora de la Concepción, 121, 126–34, 221, 229

Oxenham, John, 8, 11, 13, 16, 18, 36, 37, 117–20, 130, 224
Oxnam, John (*see* Oxenham, John)

Pacific Ocean (*see* South Sea)
Panama, xiii, 3–7, 15–17, 23–28
Panama City, 19
Papists, xx, xxvii
Parma, Duke of, 209, 212, 214
Pascha, 3–4
Pascual, Juan, 152
Payta, 137
Pedro, 16, 29–30
Pelew Islands, 180–81, 230
Pelican, 44, 47, 58, 59, 65, 75, 76, 78–79
Phillip II, King (Spain), xxi, xxv, xxx, xxxiii, xxxv, 34, 35, 39, 41, 44, 157–58, 161–65, 207–10, 224
Pike, Robert, 19–20
Pinnace, xiii
Pizarro, Francisco, xxii
Plate Fleet, xxxv–xxxvi, 14
Plymouth (England), xvi–xvii, 31, 44, 48
Point Reyes, 166
Polynesians, 181–83
Port de las Islas (*see* San Julian)
Port Desire, 63, 65
Port Pheasant, 3, 5–7
Portugal
  slave trade of, xxx
  Spain as rival of, xxvi
Portuguese
  Brazilian Indians and, 62–63
  in Spice Islands, 187–88
Potosí, xxii, 110, 112, 115, 127
Protestantism in England, xvi
Protestant Reformation, xxvi–xxvii

Quintero, 104–5

Ranse, Captain, 7
Realejo, 146
Recua, 24
Rengifo, Gomez, 159–61

Ridley, Nicholas, xx
Ridolfi Plot, 2–3
Rio de la Hacha, xxxiii–xxxiv, 221, 227
Rio de la Plata, 37
Rio de Sesto, 204
Rio Francisco, 8, 23, 28
Russian trade with England, xxviii

Sagres Castle, 211
St. Augustine (Florida), 210
St. James, 53
St. James Islands, 180
San Francisco, Cape of, 127
San Juan de Ulua, xxxv, 130, 161, 227
San Julian, 51, 65–76
San Martin, 214
San Miguel, Gulf of, 139
Santa Cruz, Marquis of, 210–12
Santa Elena, 137
Santa Marta, 5
Santiago, 53
Santiago de Tolou, 5
Santo Domingo, 210
São Pedro do Rio Grande do Sul, 60
Scilly, Isles of, 31
Sierra Leone, 204
Slavery, in Europe, xxix–xxx
Slaves, as miners at Potosí, 110–11
Slave trade, xxiv, xxx, xxxi
Smith, John, 27
Solomon Islands, 224
South America, map of, 61
South Sea, 17, 38–40, 55, 87, 94, 131–32
Southwest Passage, 35
Spain
  England and
    as ally of, xxv–xxvi
    relationship of, xxxix, 2–3, 35–36
    as rival of, xxviii
  Magellan, Strait of, expedition to, 80
  navy of, xxx
  Netherlands, revolt of against, xxviii
  New World, loss of power in, xiv
  Portugal as rival of, xxvi
  prestige, decline of, 210
  slave trade of, xxiv
  treasure of, 3
  West Indies, rumors of raid on, 48
Spaniards
  Araucanian Indians and, 95
  llamas, use of by, 113
Spanish Armada, xix, xxxix, 210–11, 212–216
Spanish Inquisition, xxvii, 107
Spanish Main, xxxiv, xxxix, 2–31
Spice Islands, 39, 46, 49, 187–92, 209
Stow, John, 2
Strait, definition of, 93, 94
Stuart, Mary, 48
Swallow, xxxii, xxxviii
Swan, 2, 4, 14, 46, 47, 60, 65

Tagoine, 204
Tanquey, bay of, 106

Tarapacá, 107–9
Tello, Rodrigo, 145
Ternate, 186–93, 221
Terra Australis, xxviii–xxix, 35, 39, 80, 86, 90, 92–94, 222
Terra Firma, 5
Tetû, Captain, 21–29, 30, 225
Thomas, Captain John, 47, 68, 73, 75
Thorne, Robert, xxvi
Tidore, 187, 188
Tierra de Guerra, 95
Tierra del Fuego, 35, 82, 84–85, 86
Tjilatjap, 201–2
Trujillo, 137
Tudor, Mary, xx, xxi, 225

"Utopia," xxvi

Valdivia, 102
Valdivia River, 95
Valparaiso, 102–7
Venta Cruz, 19, 21
Veracruz, xxxvi
Vicary, Leonard, 69

Walsingham, Sir Francis, 41, 44, 69
West Indies, xiv, xxii, xxxiii, 48, 210
William and John, xxxii
Willoughby, Hugh, xxviii
Winter, Captain John, 45–47, 55–56, 68, 71, 73, 75, 90, 134–35, 225

Zacatecas, xxii

Format by Ruth Bornschlegel
Set in Linotype Janson and Linotype Garamond
Composed by American Book–Stratford Press
Printed by Murray Printers
Bound by American Book–Stratford Press